SOLUTIONS MANUAL TO ACCOMPANY COMBINATORIAL REASONING

SOLUTIONS MANUAL TO ACCOMPANY COMBINATORIAL REASONING

An Introduction to the Art of Counting

DUANE DeTEMPLE
WILLIAM WEBB
Department of Mathematics
Washington State University
Pullman, WA

Published by John Wiley & Sons, Inc., Hoboken, New Jersey.
Published simultaneously in Canada.

Library of Congress Cataloging-in-Publication Data:
DeTemple, Duane W.
Solutions manual to accompany combinatorial reasoning : an introduction to the art of counting / Duane DeTemple, William Webb.
pages cm
ISBN 978-1-118-83078-9 (pbk.)
1. Combinatorial analysis–Textbooks. 2. Mathematical analysis–Textbooks. I. Webb, William, 1944- II. Title.
QA164.D48 2013
511′.6–dc23

2013035522

10 9 8 7 6 5 4 3 2 1

CONTENTS

PREFACE

This manual provides the statements and complete solutions to all of the odd numbered problems in the textbook *Combinatorial Reasoning: An Introduction to the Art of Counting.* The definitions, theorems, figures, and other problems referenced in the solutions contained in this manual are to that book.

HOW TO USE THIS MANUAL

The most important thing to remember is that you should not turn to any solution in this manual without first attempting to solve the problem on your own. Many of the problems are subtle or complex and therefore require considerable thought—and time!—before you can expect to find a correct method of solution. You will learn best by trying to solve a problem on your own, even if you are unsuccessful. We hope that most often you will consult this manual simply to confirm your own answer. Often your answer will be the same as ours, but sometimes you may have found a different method of solution that is not only correct, but may even be better than ours (if so, please send us your alternative solution at the address below). If the answer to a problem eludes you even after a good effort, then take a look at the solution offered here. Even in this case, it is best only to read the beginning of the solution and see if you can continue to solve the problem on your own.

TIPS FOR SOLVING COMBINATORIAL PROBLEMS

Many students wonder how to go about attacking nonroutine problems. We have listed some suggestions below that may be helpful for solving combinatorial problems and more generally for solving problems in any branch of mathematics.

- Try small cases and look for patterns
- Separate a problem into cases
- Draw a figure
- Make a table of values
- Look for a similar or related problem, one you already know how to solve
- For combinatorial problems, apply one of the strategies explored in the textbook: use the addition and multiplication principles; identify the problem as a permutation, combination, or distribution; find and solve a recurrence relation; use a generating function; use the principle of inclusion/exclusion; restate the problem to relate it to a problem answered by well known numbers such as binomial coefficients, Fibonacci numbers, Stirling numbers, partition numbers, Catalan numbers, and so on.

For a more complete discussion of mathematical problem solving, you are encouraged to consult *How to Solve It,* the classic, but still useful, book of George Pólya.

DUANE DETEMPLE
WILLIAM WEBB

Washington State University, Pullman, WA
detemple@wsu.edu and webb@math.wsu.edu

PART I

THE BASICS OF ENUMERATIVE COMBINATORICS

1

INITIAL EnCOUNTers WITH COMBINATORIAL REASONING

PROBLEM SET 1.2

1.2.1. A bag contains 7 blue, 4 red, and 9 green marbles. How many marbles must be drawn from the bag without looking to be sure that we have drawn

(a) a pair of red marbles?

(b) a pair of marbles of the same color?

(c) a pair of marbles with different colors?

(d) three marbles of the same color?

(e) a red, blue, and green marble?

Answer

(a) 18 **(b)** 4 **(c)** 10 **(d)** 7 **(e)** 17

1.2.3. There are 10 people at a dinner party. Show that at least two people have the same number of acquaintances at the party.

Answer

Each person can know any where from 0 (no one) to 9 (everyone) people. But if someone knows no one, there cannot be someone who knows

Solutions Manual to Accompany Combinatorial Reasoning: An Introduction to the Art of Counting, First Edition. Duane DeTemple and William Webb.

everyone, and vice versa. Thus, place the 10 people into the 9 boxes that are labeled 1, 2, … , 8, and 0|9. By the pigeonhole principle, some box has at least 2 members. That is, there are at least two people at the party with the same number of acquaintances.

1.2.5. Given any five points in the plane, with no three on the same line, show that there exists a subset of four of the points that form a convex quadrilateral.

[*Hint*: Consider the *convex hull* of the points; that is, consider the convex polygon with vertices at some or all of the given points that encloses all five points. This scenario can be imagined as the figure obtained by bundling the points within a taut rubber band that has been snapped around all five points. There are then three cases to consider, depending on whether the convex hull is a pentagon, a quadrilateral containing the fifth point, or a triangle containing the other two given points.]

Answer
If the convex hull is a pentagon, each set of 4 points are the vertices of a convex quadrilateral. If the convex hull is a quadrilateral, the convex hull itself is the sought quadrilateral. If the convex hull is a triangle, the line formed by the two points within the triangle separates the vertices of the triangle into opposite half planes. By the pigeonhole principle, there are two points of the triangle in the same half plane. These two points, together with the two points within the triangle, can be combined to form the desired convex quadrilateral.

1.2.7. Given five points on a sphere, show that some four of the points lie in a closed hemisphere.

[*Note*: A closed hemisphere includes the points on the bounding great circle.]

Answer
Pick any two of the five points and draw a great circle through them. At least two of the remaining three points belong to the same closed hemisphere determined by the great circle. These two points, and the two starting points, are four points in the same closed hemisphere.

1.2.9. Suppose that 51 numbers are chosen randomly from $[100] = \{1, 2, \dots, 100\}$. Show that two of the numbers have the sum 101.

Answer
Each of the 51 numbers belongs to one of the 50 sets $\{1, 100\}, \{2, 99\}, \dots, \{50, 51\}$. Some set contains two of the chosen numbers, and these sum to 101.

1.2.11. Choose any 51 numbers from $[100] = \{1, 2, \dots, 100\}$. Show that there are two of the chosen numbers that are relatively prime (i.e., have no common divisor other than 1).

Answer
Place each of the 51 numbers into one of the 50 sets $\{1, 2\}, \{3, 4\}, \dots,$ $\{99, 100\}$. One of the sets contains a pair of consecutive integers that are relatively prime.

1.2.13. Choose any 51 numbers from $[100] = \{1, 2, \dots, 100\}$. Show that there are two of the chosen numbers for which one divides the other.

Answer
Any natural number has the form $m = 2^{d_m} k_m$, where $d_m \geq 0$ and k_m is odd. Call k_m the *odd factor* of m. For example, the odd factor of $100 = 2^2 \cdot 25$ is $k_{100} = 25$. Thus, the odd factors of the 51 chosen numbers are in the set $\{1, 3, 5, \dots, 99\}$. Since this is a set with 50 members, two of the 51 chosen numbers have the same odd factor. The smaller is then a divisor of the larger, with a quotient that is a power of two.

1.2.15. Consider a string of $3n$ consecutive natural numbers. Show that any subset of $n + 1$ of the numbers has two members that differ by at most 2.

Answer
Suppose the $3n$ consecutive numbers are $a, a + 1, \dots, b$. Each of the $n + 1$ numbers in the given subset belongs to one of the sets $\{a, a + 1, a + 2\}$, $\{a + 3, a + 4, a + 5\}. \dots, \{b - 2, b - 1, b\}$. By the pigeonhole principle, one of these sets has two members of the subset and these differ by at most 2.

1.2.17. Suppose that the numbering of the squares along the spiral path shown in Example 1.9 is continued. What number k is assigned to the square S whose lower left corner is at the point $(9, 5)$?

Answer
We want to find a solution to the equations $k = 11i + 9$ and $k = 16j + 5$ for some integers i and j. This gives us $11i + 4 = 16j$. Both 4 and 16 are divisible by 4, so we see that i is divisible by 4. If we let $i = 4$, then $j = 3$ and we obtain the solution $k = 53$. The next multiple of 4 giving a solution is $i = 20$, but then $k = 229$ and we see that the spiral is overlapping itself with repeated squares covered a second time.

1.2.19. Generalize the results of Problem 1.2.18.

(a) How many spiral paths exist on the torus if $m = n$?

(b) Suppose $d \geq 2$ is the largest common divisor of m and n. How many distinct spiral paths exist on the torus?

Answer

(a) Any path returns to its starting position in m steps, so there are m spirals each covering m squares. For example, there are three paths when $m = n = 3$, as shown here.

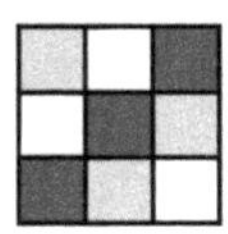

(b) Since m/d and n/d are relatively prime, there is a unique spiral with mn/d^2 steps that covers a d by d square at each step. For example, if $m = 6$ and $n = 9$, then $d = 3$, and there is a unique spiral of length $\frac{mn}{d^2} = \frac{6 \cdot 9}{3^2} = 6$ of 3×3 squares that covers the torus. This is seen at the left in the figure below. By part (a) we see there are d nonintersecting spirals on the torus, each of length $\frac{mn}{d}$. The case $m = 6$, $n = 9$, $d = 3$, is shown at the right below, with the $d = 3$ paths each of length $\frac{mn}{d} = \frac{6 \cdot 9}{3} = 18$ shown in black, white, and gray.

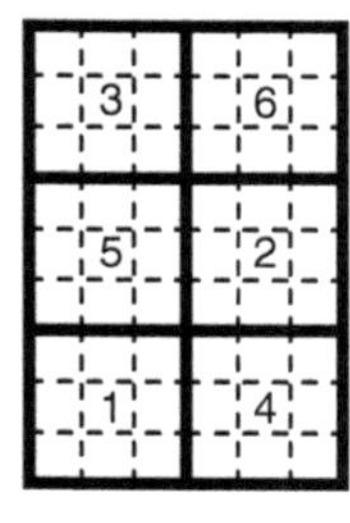

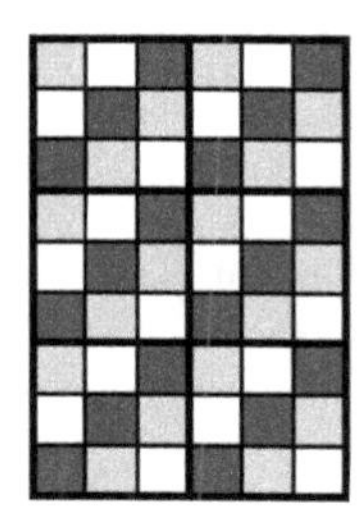

PROBLEM SET 1.3

1.3.1. Consider an $m \times n$ chessboard, where m is even and n is odd. Prove that if two opposite corners of the board are removed, the trimmed board can be tiled with dominoes.

Answer

The left and right hand columns of height $n - 1$ of the trimmed board can each be tiled with vertical dominoes. The remaining board is has all of its rows of even length $m - 2$, so it can be tiled with horizontal dominoes.

1.3.3. Suppose that the lower left $j \times k$ rectangle is removed from an $m \times n$ chessboard, leaving an angle-shaped chessboard. Prove that that angular board can be tiled with dominoes if it contains an even number of squares.

Answer

Since $mn - jk = (m - k)n + (n - j)k$ is even, $(m - k)n$ and $(n - j)k$ have the same parity. If both are even, we can tile the resulting $(m - k) \times n$ and $(n - j) \times k$ rectangles. If both are odd, then n and k are odd thus m and j must be even. We can then tile the $m \times (n - j)n$ and $(m - k) \times j$ rectangles.

Alternate answer

View the angular region as the union of rectangles A, B, and C, where the corner rectangle B shares an edge with each of A and C. If all three rectangles have even area, the angle can be tiled since A, B, and C can each be tiled individually. If A and B, or B and C, each have odd area, then combining the odd rectangles shows that the angle is a union of two even area rectangles and therefore can be tiled. If A and C are odd, their edges are all of odd length and therefore rectangle B is also odd; the angular board therefore is not of even area.

1.3.5. Consider a rectangular solid of size $l \times m \times n$, where l, m, and n are all odd positive integers. Imagine that the unit cubes forming the solid are alternately colored gray and black, with a black cube at the corner in the first column, first row, and first layer.

(a) What is the color of each of the remaining corner cubes of the solid?

(b) How can the color of the cube in column j, row k, and layer h of the solid be determined?

(c) Prove that removing any black cube leaves a trimmed solid that can be filled with solid $1 \times 1 \times 2$ dominoes.

Answer

(a) Since the colors alternate, all eight corners of the solid are black.

(b) The cube is black if and only if the sum $j + k + h$ is odd. For example, the cube in column 1, row 1, and layer 1 is black since $1 + 1 + 1 = 3$, an odd number. That is, j, k, and h must all be odd, or one must be odd and the other two even.

(c) If the cube that is removed is black, and is in column j, row k, and layer h, then j, k, and h are all odd or two are even and one is odd. With no loss in generality assume that $j + k$ is even and h is odd. Theorem 1.21 tells us that layer h can be tiled with dominoes confined to that layer. When layer h is removed it leaves two (possibly one if $h = 1$ or n) rectangular solids with an even dimension and so it can be tiled with solid dominoes.

1.3.7. A *tetromino* is formed with four squares joined along common edges. For example, the O and the Z tetromino are shown here.

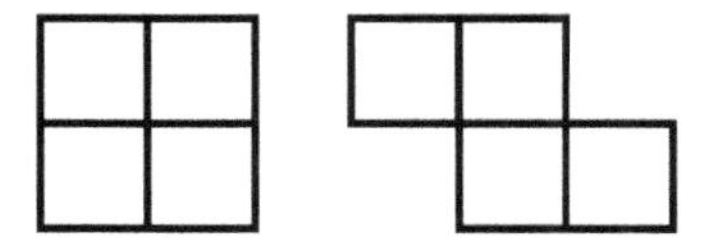

(a) Find the three other tetrominoes, called the I, J, and T tetrominoes.

(b) The set of five tetrominoes has a total area of 20 square units. Explain why it is not possible to tile a 4×5 rectangle with a set of tetrominoes.

(c) Show that a 4×10 rectangle can be tiled with two sets of tetrominoes.

(d) Show that a 5×8 rectangle can be tiled with two sets of tetrominoes.

Answer

(a)

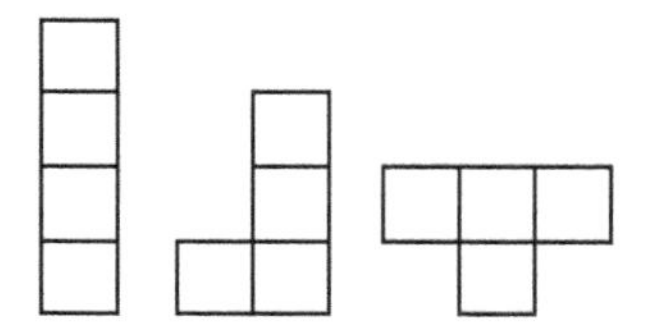

(b) A 4×5 chessboard has 10 unit squares of each color. The O, Z, I, and J tetrominoes each cover 2 unit squares of each color, but the T tetromino covers 3 squares of one color and one of the other color. Therefore the 4×5 square cannot be tiled with a set of tetrominoes.

(c)

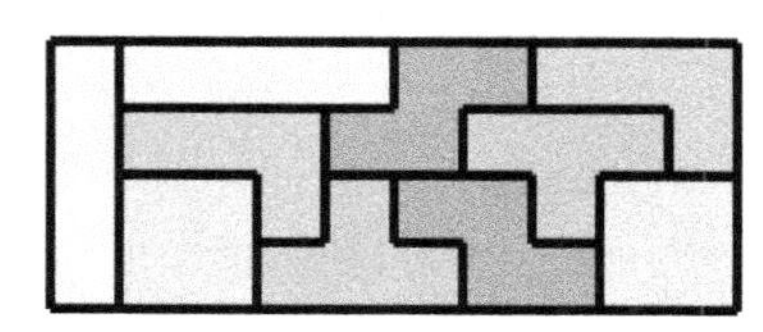

(d)

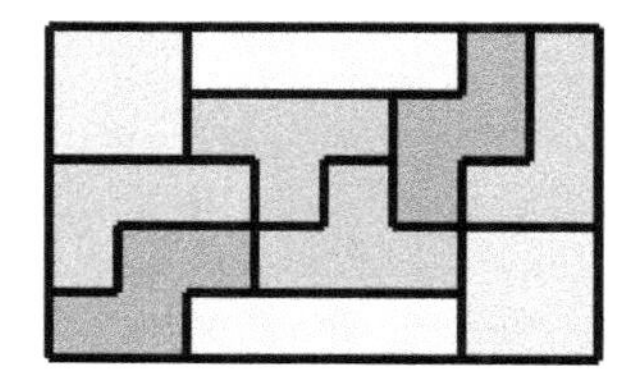

PROBLEM SET 1.4

1.4.1. The following diagram illustrates that $t_{2m} = m(2m+1)$

Create a similar diagram that illustrates the formula $t_{2m+1} = (2m+1)(m+1)$.

Answer

1.4.3. Use both algebra and dot patterns to show that the square of an odd integer is congruent to 1 modulo 8. That is, show that $s_{2n+1} = 8u_n + 1$ for some integer u_n. Be sure to identify the integer u_n by its well-known name.

Answer

The answer is $s_{2n+1} = 8t_n + 1$, since $(2n+1)^2 = 4n^2 + 4n + 1 = 8\frac{n(n+1)}{2} + 1 = 8t_n + 1$. See the following diagram:

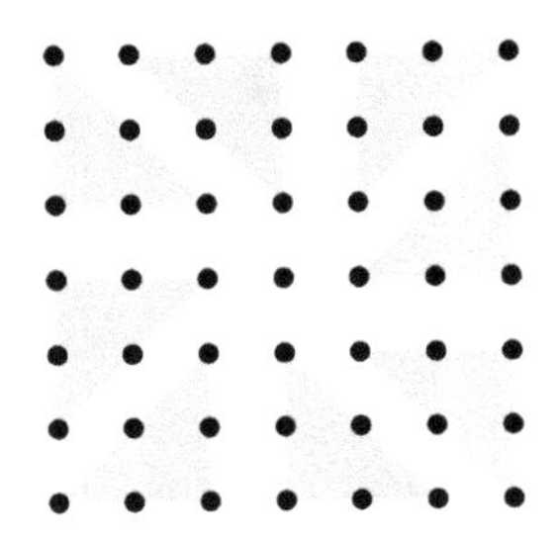

1.4.5. The *centered square numbers* are obtained much like the centered triangle numbers of Problem 1.4.4, except that squares with an increasing number of dots per side surround a center dot.

(a) Create a diagram that shows the sequence of centered square numbers beginning with 1, 5, 13, 25, and 41.

(b) Color the dots in the diagram from part (a) to show that the nth centered square number is given by $(n+1)^2 + n^2$.

(c) Shade your diagram from part (a) to shows that every centered square number is congruent to 1 modulo 4.

(d) Verify part (c) with algebra.

Answer

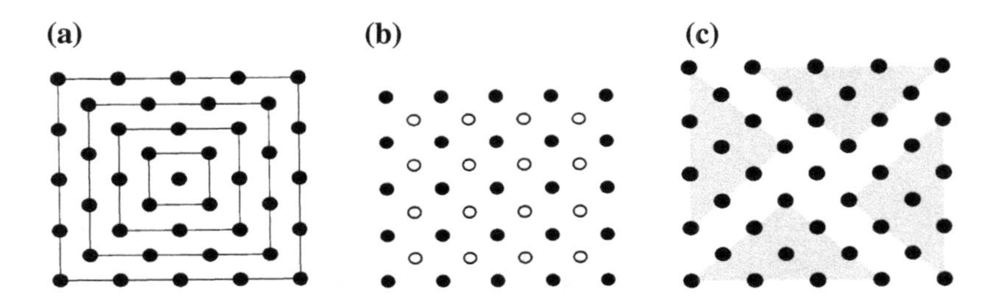

(d) $(n+1)^2 + n^2 = 2n^2 + 2n + 1 = 1 + 2n(n+1) = 1 + 4\frac{n(n+1)}{2} = 1 + 4t_n$

1.4.7. The first three *trapezoidal numbers* are 1, 5, and 12, as shown by the dot pattern here.

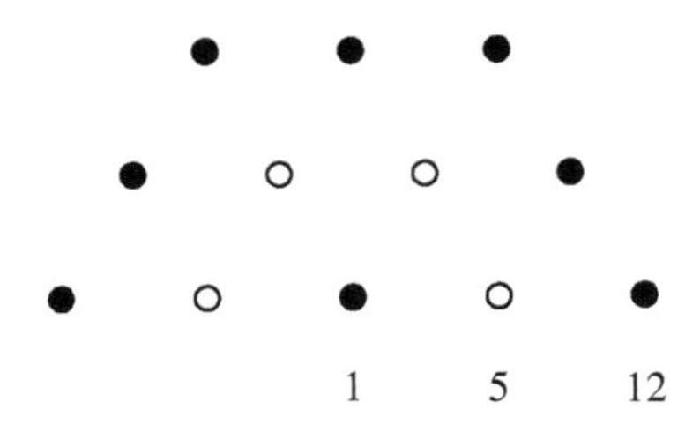

(a) Continue the trapezoidal pattern to find the next three trapezoidal numbers.

(b) Draw some lines on your diagram from part (a) to explain why the trapezoidal numbers are simply an alternative pattern for the pentagonal numbers $p_1 = 1, p_2 = 5, p_3 = 12, \ldots$.

(c) Use the trapezoidal diagram to show why each pentagonal number is the sum of a triangular number and a square number. Give an explicit formula for p_n in terms of the triangular and square numbers.

(d) The trapezoidal diagram shows that each pentagonal number is the difference of two triangular numbers. Determine the two triangular numbers corresponding to p_n and express this result in a formula.

(e) Construct a diagram showing that each pentagonal number is one-third of a triangular number. Give an explicit formula of this property.

Answer

(a) The next three trapezoidal numbers are 22, 35, and 51.

(b) View a trapezoidal number as a distorted pentagon *ABCDE,* with sides *EA* and *AB* along the same line.

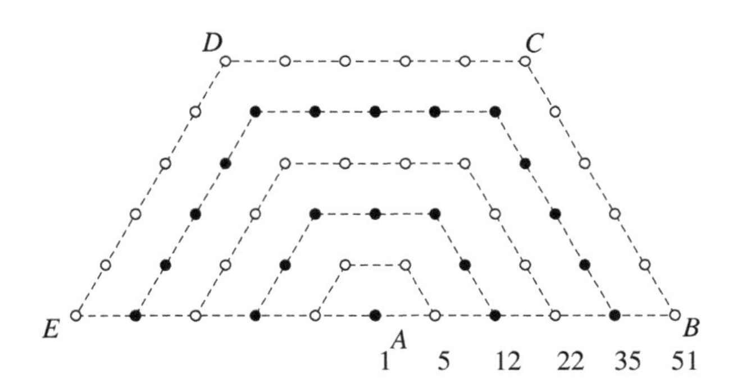

(c) $p_n = t_n + n^2$. For example, $p_6 = 51 = t_{6-1} + 6^2 = \dfrac{5 \cdot 6}{2} + 36 = 15 + 36$

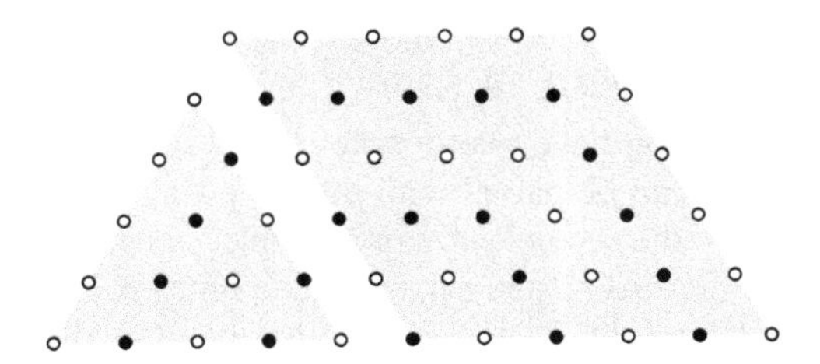

(d) Extend the trapezoid to a triangle of side $2n-1$. We then see that $p_n = t_{2n-1} - t_{n-1}$. For example, $p_4 = 22 = t_7 - t_3 = \frac{7 \cdot 8}{2} - \frac{3 \cdot 4}{2} = 28 - 6$.

(e) The shading in the diagram below shows that $p_n = \frac{1}{3}t_{3n-1}$.

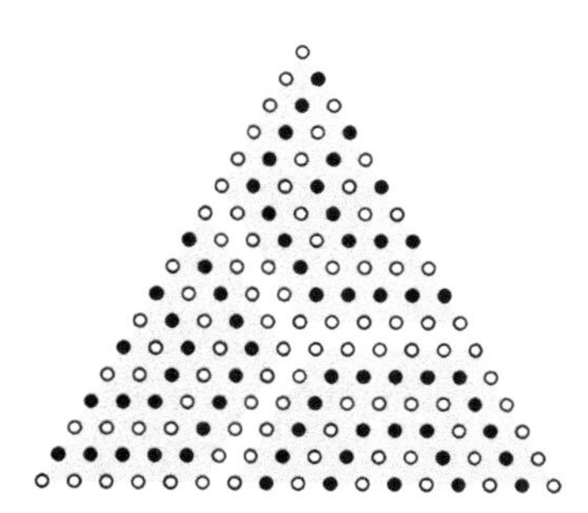

For example, $p_6 = 51 = \frac{1}{3}t_{3 \cdot 6-1} = \frac{1}{3}t_{17} = \frac{1}{3}\left(\frac{17 \cdot 18}{2}\right) = \frac{17 \cdot 18}{6} = 17.3$.

1.4.9. Dominoes, as described in Problem 1.4.8 also come in double-9, double-12, double-15, and even double-18 sets. Consider, more generally, a double-n set, so each half-domino is imprinted with 0 to n pips.

(a) Derive a formula for the number of dominoes in a double-n set. Use the formula to determine the number of dominoes in a double-n set for $n = 6, 9, 12, 15$, and 18.

(b) Derive a formula for the total number of pips in a double-n set. Use the formula to determine the total number of pips in a double-n set for $n = 6, 9, 12, 15$, and 18.

Answer

(a) Consider the array of dots in the x-,y-coordinate plane with a dot at (p, q) that represents the $p - q$ domino, with $0 \leq q \leq p \leq n$. This array is a triangle with $n + 1$ dots per side, so there are $t_{n+1} = \frac{1}{2}(n+1)(n+2)$ dominoes in a double-n set.

For $n = 3, 6, 9, 12, 15$, and 18, the number of dominoes are the triangular numbers 10, 28, 55, 91, 136, and 190.

(b) Imagine that you have two double-n sets, so that each $p - q$ domino from one set can be paired with the $(n-p)-(n-q)$ complementary domino from the second set. For example, in a double-15 set, pair the $11 - 6$ domino from one set with the complementary $4 - 9$ domino from the second set. Each pair of complementary dominoes has a total of $2n$ pips, so by part (a) there are $(2n)t_{n+1} = (2n)\frac{1}{2}(n+1)(n+2) = n(n+1)(n+2)$ pips in the two double-n sets. Therefore, a single double-n set has a total of $\frac{1}{2}n(n+1)(n+2)$ pips.

Alternate answer

Each half-domino with k pips, $0 \le k \le n$, occurs $n+2$ times in a double-n set, so the total number of pips is given by $(n+2)(0+1+2+\cdots+n) = (n+2)t_n = (n+2)\frac{n(n+1)}{2} = \frac{1}{2}n(n+1)(n+2)$.

PROBLEM SET 1.5

1.5.1. **(a)** Extend Figure 1.9 to depict the set of 16 tilings of a board of length 4, where each tile is either gray or white.

(b) Explain how it is easy to use the 8 tilings of boards of length 3 to draw all of the tiled boards of length 4.

Answer

(a)

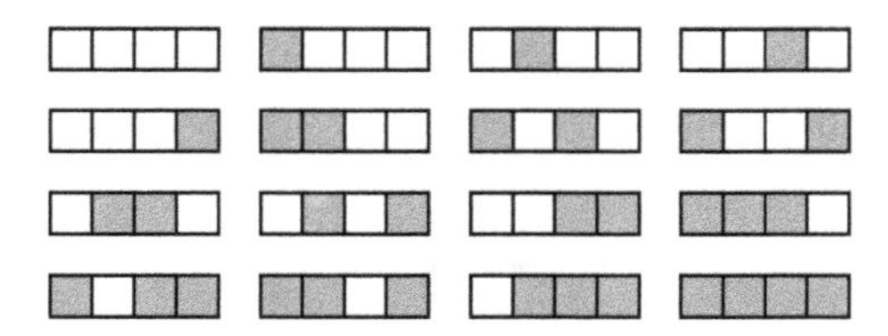

(b) Add a white tile at the right end of each of the 8 tiled boards of length 3, and then add a gray tile at the right end of each of the 8 tiled boards of length 3. Altogether, this forms all 16 of the tilings of boards of length 4.

1.5.3. Use formulas (1.20) and (1.21) to prove Pascal's identity (1.24).

Answer

$$\binom{n-1}{k-1} + \binom{n-1}{k} = \frac{(n-1)!}{(k-1)!\,(n-k)!} + \frac{(n-1)!}{k!\,(n-k-1)!}$$
$$= \frac{(n-1)!}{k!\,(n-k)!}[k+n-k] = \frac{n!}{k!\,(n-k)!} = \binom{n}{k}$$

1.5.5. **(a)** Find all of the ways that a 2×4 rectangular board can be tiled with 1×2 dominoes. Here is one way to tile the board.

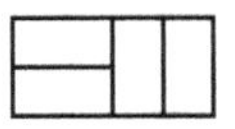

(b) Draw all of the ways to tile a 2×4 board with dominoes.

(c) How many ways can a $2 \times n$ board be tiled with dominoes?

Answer

(a) There are five tilings:

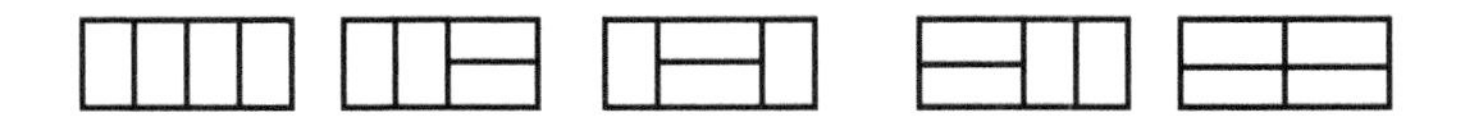

(b) Draw a horizontal midline through each tiling by dominoes. The lower $1 \times n$ row board is equivalent to a tiling by squares and dominoes, showing there are f_n tilings of a $2 \times n$ board with dominoes.

1.5.7. The following train (see Problem 1.5.6 for the definition of a train) has just one car of length 13.

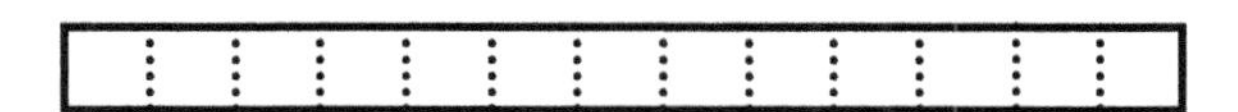

(a) How many ways can a train of length 13 be formed with 2 cars?

(b) Why are there $\binom{12}{4}$ trains of length 13 that can be formed with 5 cars?

(c) Generalize your answer to part (b) to give a binomial coefficient that expresses the number of trains of length n with r cars.

Answer

(a) Any one of the 12 vertical dashed lines can be chosen to end a car and start a new one, so there are $\binom{12}{1} = 12$ ways to form a train of length 13 with 2 cars.

(b) Any choice of 4 of the 12 vertical dashed lines forms a train with 5 cars. Thus, there are $\binom{12}{4} = \frac{12 \cdot 11 \cdot 10 \cdot 9}{4 \cdot 3 \cdot 2 \cdot 1} = 495$ trains of length 13 with 5 cars.

(c) There are $\binom{n-1}{r-1}$ trains of length n with r cars.

1.5.9. There are 4 ways to express 5 as a sum of two ordered summands, namely $4 + 1$, $3 + 2$, $2 + 3$, and $1 + 4$.

(a) How many ways can 5 be expressed as a sum of three ordered summands? (see Problem 1.5.8)

(b) How many ways can a positive integer n be expressed as a sum of k summands?

Answer

(a) Six ways

(b) $\binom{n-1}{k-1}$ since the summands can be viewed as the lengths of k cars that form a train of length n.

1.5.11. How many binary sequences of length n have no two consecutive ones? (A binary sequence in an ordered list of ones and zeroes, such as 100101001.) For example, there are 5 binary sequences of length 3 with no two consecutive ones, namely 000, 100, 010, 001, and 101.

Answer

There are 2 sequences of length 1 (0 and 1), 3 of length 2 (00, 10, and 01), 5 of length 3, and 8 of length 4 (0000, 1000, 0100, 0010, 0001, 1010, 0101, and 1001). This suggests that the number of binary sequences of length n with no consecutive ones is given by the Fibonacci number F_{n+2}, or, equivalently, by the combinatorial Fibonacci number f_{n+1}. But this is the number of tilings of a $1 \times (n + 1)$ board with squares and dominoes. To see the connection, consider a board of length 4. Each of the three vertical dashed segments can be labeled with a 0 to represent a break between tiles or labeled with a 1 to represent the midline of a domino. Since the distance between midlines of any two dominos is at least two, no two 1s can be consecutive. Here is the correspondence between tilings and binary sequences with no consecutive ones in the case of a sequence of length 3 and a board of length 4:

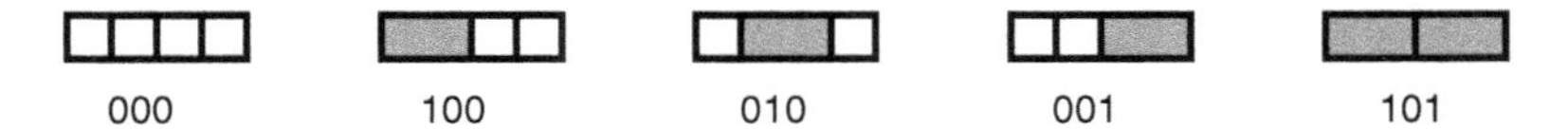

PROBLEM SET 1.6

1.6.1. The 2006 World Cup used a soccer ball called the "+teamgeist" ("team spirit," with the + sign to allow the German word to be copyrighted) Mathematically, the ball is a spherical analog of the truncated octahedron, obtained by starting with the octahedron of 8 triangles as shown, then replacing ("truncating") each corner with a square, and finally rounding the faces to become spherical. What counting principle can be used to determine the number of panels on the +teamgeist soccer ball?

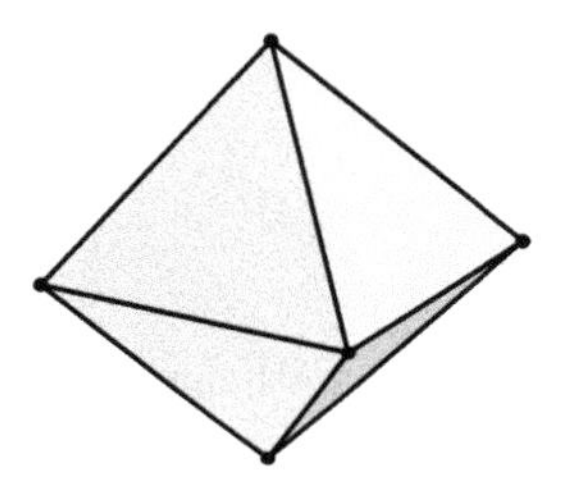

Answer

When truncated, the 6 corners of the octahedron are replaced with squares, and the 8 triangles become hexagons. The sets of squares and hexagons partition the set of all of the ball's panels, so there are $8 + 6 = 14$ panels in all by the addition principle.

1.6.3. There are two types of seams of the traditional soccer ball shown in Example 1.30, those that separate two hexagonal panels and those that separate a pentagonal panel from a hexagonal panel. Find the total number of seams on the ball, and the number that separate one hexagonal panel from another.

Answer

Let S denote the set of all of the seams on the ball, A the set of seams between panels of different types, and B the set of seams between abutting hexagonal panels. Since every seam in set A borders exactly one of the 12 pentagons, we see that $|A| = 5 \cdot 12 = 60$. Similarly, each of the 20 hexagonal panels is bordered by 6 seams, but the product $6 \cdot 20 = 120$ counts the seams between abutting hexagons twice and the seams along pentagons once. This means that $120 + 60 = 180$ counts every seam of the ball twice, so we conclude there are $|S| = 180/2 = 90$ seams on the ball. By the subtraction principle, $90 - 60 = 30$ seams separate the hexagonal panels from one another.

1.6.5. Maria likes to order double-scoop ice cream cones, with chocolate or strawberry on the bottom, and chocolate, vanilla, or mint on the top. Describe the ways Maria can order her ice cream cones with a Cartesian product, and count the number of types of cones she likes.

Answer

Let $A = \{C, S\}$ and $B = \{C, V, M\}$ be the sets of choices of flavors. Then Maria will have $|A \times B| = |A||B| = 2 \cdot 3 = 6$ types of cones to her liking.

1.6.7. Generalize the results of Example 1.45 and Problem 1.6.6. That is, provide formulas for the number of ways to split up m people into singles and doubles matches. For doubles, create as many matches as possible and then set up a singles match if enough people remain not already playing doubles.

Answer

If $m = 2n$, we can form $(2n - 1)!!$ singles matches, and if $m = 2n + 1$ is odd we can form n matches in $(2n + 1)!!$ ways, with someone sitting out. If $m = 4n$, we can form $2n$ doubles partners in $(4n - 1)!!$ ways, and then form doubles matches in $(2n - 1)!!$ ways. That is, there are $(4n - 1)!!(2n - 1)!!$ ways to split up for n games of doubles. If $m = 4n + 2$, we can form $2n + 1$ pairings in $(4n + 1)!!$ ways, and then form doubles matches and one singles match in $(2n + 1)!!$ ways. That is, there are $(4n + 1)!!(2n + 1)!!$ ways to split up for n doubles and one singles match. If $m = 4n + 1$, we can choose a player to sit out in $4n + 1$ ways, so there are $(4n + 1)(4n - 1)!!(2n - 1)!! = (4n + 1)!!(2n - 1)!!$ ways to form the n doubles matches and choose who sits out. Similarly, for If $m = 4n + 3$, we can choose a player to sit out in $4n + 3$ ways, so there are $(4n + 3)(4n + 1)!!(2n + 1)!! = (4n + 3)!!(2n + 1)!!$ ways to form the n doubles matches, one singles match, and choose who sits out.

1.6.9. Construct Pascal's triangle that shows rows 0 through 6.

(a) Draw a loop around the terms in the hockey stick identity (1.32) for the case $n = 3$ and $r = 2$, showing that a hockey stick shape is formed.

(b) Repeat part (a) but for the hockey stick identity of Problem 1.6.8 in the case that $n = 5$ and $r = 3$.

(c) How do the terms in the handle of the hockey stick relate to the term in the blade of the hockey stick?

Answer

(a) and **(b)**: See the diagram below.

(c) The sum of the terms in the handle is the term in the blade.

$$
\begin{array}{ccccccccccccc}
 & & & & & & 1 & & & & & & \\
 & & & & & 1 & & 1 & & & & & \\
 & & & & 1 & & 2 & & 1 & & & & \\
 & & & 1 & & 3 & & 3 & & 1 & & & \\
 & & 1 & & 4 & & 6 & & 4 & & 1 & & \\
 & 1 & & 5 & & 10 & & 10 & & 5 & & 1 & \\
1 & & 6 & & 15 & & 20 & & 15 & & 6 & & 1
\end{array}
$$

1.6.11. There are three roads from Sylvan to Tacoma, four roads from Tacoma to Umpqua, and two roads from Sylvan directly to Umpqua. How many routes, with no backtracking, can be taken from Sylvan to Umpqua?

Answer

$3 \cdot 4 + 2 = 14$

SECTION 1.7

1.7.1. **(a)** Let A be any set of 51 numbers chosen from [100]. Show that two members of A differ by 50.

(b) State and prove a generalization of the result of part (a).

Answer

(a) Consider the 50 pigeonholes given by the sets $\{1, 51\}$, $\{2, 52\}, \dots,$ $\{50, 100\}$. By the pigeonhole principle, two of the 51 numbers belong to the same set and their difference is 50.

(b) Theorem. Any subset of $[2n]$ with $n + 1$ members contains two members whose difference in n.

Proof. Consider the n pigeonholes $\{1, n + 1\}$,$\{2, n + 2\}, \dots,$ $\{n, 2n\}$. At least two members of any set with $n + 1$ members has 2 members in the same pigeonhole, and their difference is n.

1.7.3. Show that any set of 10 natural numbers, each between 1 and 100, contains two disjoint subsets with the same sum of its members.

Answer

There are $2^{10} = 1024$ subsets of a set with 10 members. The sum of the numbers in any subset less than $10 \cdot 100 = 1000$. By the pigeonhole principle, there are at least two distinct subsets with the same sum. If there are any numbers common to both sets, these can be deleted from both sets to leave two disjoint sets still with the same sum.

1.7.5. A traditional dart board divides the circular board into 20 sectors that are numbered clockwise from the top with the sequence 20 - 1 - 18 - 4 - 13 - 6 - 10 - 15 - 2 - 17 - 3 - 19 - 7 - 16 - 8 - 11 - 14 - 9 - 12 - 5.

There is considerable variation in the sum of three successive numbers, from $23 = 1 + 18 + 4$ to $42 = 19 + 7 + 16$. Can the numbers 1 through 20 be rearranged so that the sum of each group of three successive numbers is smaller than 32?

Answer

Consider an arrangement $a_1, a_2, \dots, a_{20}$ and the 20 sums $s_1 = a_1 + a_2 + a_3, s_2 = a_2 + a_3 + a_4, \dots, s_{20} = a_{20} + a_1 + a_2$. Then $s_1 + s_2 + \cdots + s_{20} = 3\left(a_1 + a_2 + \cdots + a_{20}\right) = 3(1 + 2 + \cdots + 20) = 3\frac{20 \cdot 21}{2} = 630$.

If each sum were no larger than 31, then $s_1 + s_2 + \cdots + s_{20} \leq 20 \cdot 31 = 620$. This is a contradiction, so at least one of the sums s_j must be 32 or larger.

Alternate answer

The average of the sums $s_1, s_2, \dots, s_{20}$ is $\frac{s_1 + s_2 + \cdots + s_{20}}{20} = \frac{3\left(a_1 + a_2 + \cdots + a_{20}\right)}{20} = \frac{3(1 + 2 + \cdots + 20)}{20} = 3\frac{20 \cdot 21}{2 \cdot 20} = 31.5$. If each sum were no larger than 31, then their average would also be no larger than 31, a contradiction.

1.7.7. The centered hexagon numbers (or *hex* numbers), H_n, are obtained by starting with a single dot and then surrounding it by hexagons with 6, 12, 18, ... dots on its sides. The diagram below shows that $H_0 = 1$, $H_1 = 7$, and $H_2 = 19$.

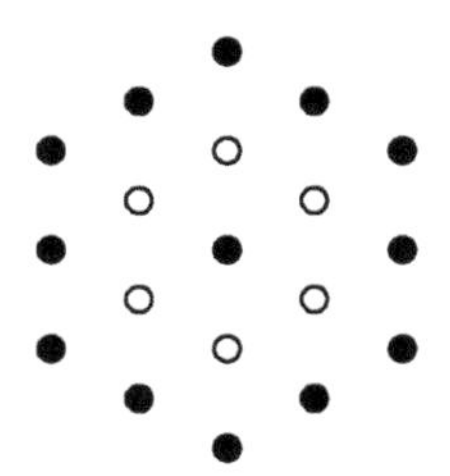

(a) Extend the diagram with two more surrounding hexagons to determine H_3 and H_4.

(b) Derive a formula that gives H_n in terms of the triangular numbers.

(c) Obtain an expression for H_n as a function of n.

(d) Suppose that *tridominoes* are formed with a pair of equilateral triangles joined along a common edge, and are colored gray, white, or black according to their orientation. The figure here shows the three types of tridominoes and one way they can be used to tile the hex pattern for the H_2 array of dots.

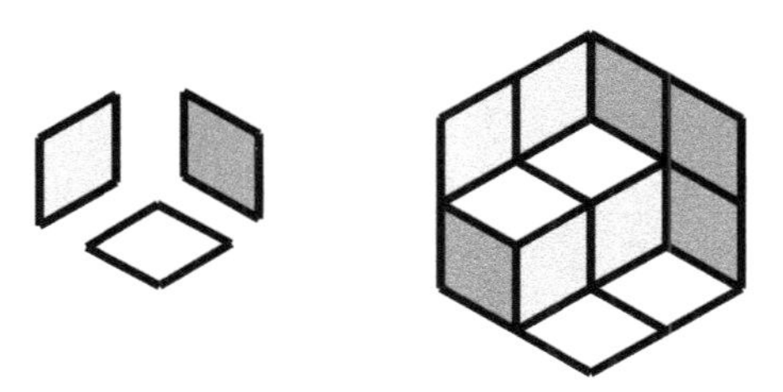

Show that every hex pattern of H_n dots can be tiled with tridominoes, and give the number of tridominoes of each color that are used in the tiling. [*Suggestion*: The hex numbers might also be called the *corner* numbers!]

Answer

(a) $H_3 = 37, H_4 = 61$

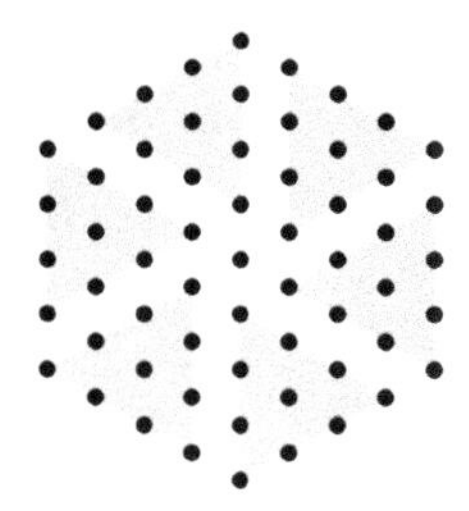

(b) The shading of the diagram shows that $H_n = 1 + 6t_n$.

(c) $H_n = 1 + 6t_n = 1 + 6\left(\frac{1}{2}n(n+1)\right) = 3n^2 + 3n + 1$

(d) The tiling with colored tridominoes makes it appear that cubes have been stacked into a cubical $n \times n \times n$ corner. When viewed from above, only the n^2 white tridominoes show, and similarly n^2 black and n^2 gray tridominoes are seen from the right and left, respectively. Altogether, $3n^2$ tridominoes are needed for the tiling. For example, if $n = 7$, there are 49 tridominoes of each color, as seen in the tiling that follows.

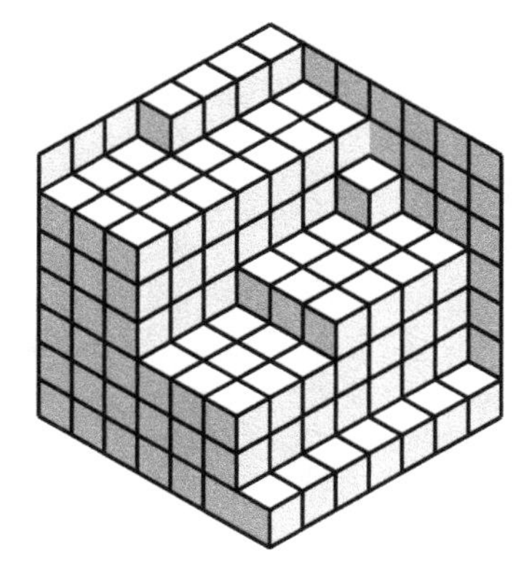

1.7.9. Use the result of Problem 1.7.8 to show that hockey stick identity 1 (1.32) can be rewritten to become $\binom{n+1}{r+1} = \binom{r}{r} + \binom{r+1}{r} + \binom{r+2}{r} + \cdots + \binom{n}{r}$ (hockey stick identity 2)

Answer

Hockey stick identity 1 (1.32), with n replaced by s and r replaced by $m - s$, becomes $\binom{m+1}{m-s} = \binom{s}{0} + \binom{s+1}{1} + \binom{s+2}{2} + \cdots + \binom{m}{m-s}$.

Using the result $\binom{n}{r} = \binom{n}{n-r}$ from Problem 1.7.8, this identity becomes $\binom{m+1}{s+1} = \binom{s}{s} + \binom{s+1}{s} + \binom{s+2}{s} + \cdots + \binom{m}{s}$.

This is hockey stick identity 2 once s is replaced with r and m is replaced n.

1.7.11. **(a)** How many paths extend from point P_1 to each of the points $P_2, P_3, \ldots, P_{12}$ in the following directed graph? Each step along any path must be in the direction indicated by the arrow. For example, there are 2 paths from P_1 to P_3.

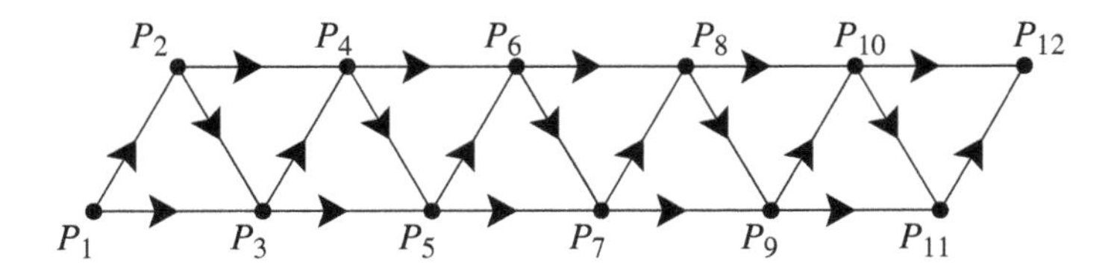

What famous number sequence gives the number of paths to the points $P_1, P_2, \ldots, P_n$? Provide a justification for your answer.

Answer

(a) There are 1, 2, 3, 5, 8, 13, 21, 34, 55, 89, and 144 paths, respectively, to the points $P_2, P_3, \ldots, P_{12}$.

(b) The set of paths that reach point P_n can be partitioned into two disjoints subsets: the set of paths that end crossing the arc $P_{n-1}P_n$ and the set of paths that end by crossing the arc $P_{n-2}P_n$. This partition shows that the number of paths to vertex P_n is the sum of the number of paths to vertices P_{n-1} and P_{n-2}. This is the Fibonacci recursion relation, and there is one path to P_2 and two paths to P_3. Therefore, the number of paths to points P_n is the Fibonacci number F_n.

1.7.13. There are six positive divisors of $12 = 2^2 \times 3^1$, namely $1 = 2^0 \times 3^0$, $2 = 2^1 \times 3^0$, $4 = 2^2 \times 3^0$, $3 = 2^0 \times 3^1$, $6 = 2^1 \times 3^1$, and $12 = 2^2 \times 3^1$. What is the number of positive divisors of these integers?

(a) 660 **(b)** $2^5 \times 3^7 \times 11^2 \times 23^4$ **(c)** 10^{100}

Answer

(a) $660 = 2^2 \times 3^1 \times 5^1 \times 11^1$, so there are $3 \times 2 \times 2 \times 2 = 24$ positive divisors, since we can choose the powers of 2, 3, 5, and 11 in 3, 2, 2, and 2 ways, respectively in each prime divisor.

(b) $6 \times 8 \times 3 \times 5 = 720$ positive divisors

(c) $10^{100} = 2^{100} \times 5^{100}$ has $101^2 = 10201$ divisors

2

SELECTIONS, ARRANGEMENTS, AND DISTRIBUTIONS

PROBLEMS SET 2.2

Unless the problem specifically asks for a numerical evaluation, it is enough to answer with an expression involving factorials, permutations, and/or combinations. However, your solution must clearly explain the reasoning that you have used to obtain the expression. When asked to use "combinatorial reasoning," base your reasoning on the combinatorial meaning of combinations, permutations, and the like. In particular, avoid algebraic calculation and numerical calculation. To give a combinatorial proof of an identity, propose a counting question and then answer it in two ways, so that equating the expressions derived in your two answers yields the identity.

2.2.1. What is the number of permutations of the 26 letters of the alphabet

(a) with no restrictions?

(b) with all 5 vowels in the consecutive group *aeiou*?

(c) with all 5 vowels in a group of 5, though in any order within the group?

(d) with the 5 vowels appearing in the natural order *a, e, i, o, u,* though not necessarily in a group of consecutive letters?

Solutions Manual to Accompany Combinatorial Reasoning: An Introduction to the Art of Counting, First Edition. Duane DeTemple and William Webb.

Answer

(a) 26!

(b) 22!

(c) $5! \cdot 22!$

(d) $\binom{26}{5} 21!$

2.2.3. What is the number of permutations of the digits 0, 1, 2, … , 9

(a) with no restrictions?

(b) if the odd digits are adjacent to one another?

(c) if no two odd digits are adjacent?

(d) if every prime is to the left of every nonprime?

Answer

(a) 10!

(b) $6 \cdot (5!)^2$

(c) $6 \cdot (5!)^2$ since the even and odd digits can each be permuted in 5! ways and there are 4 permutations beginning and ending with an odd, and 2 that either begin or end with an even

(d) $4! \cdot 6!$

2.2.5. There are three flagpoles and 11 different flags, where each pole can fly up to 4 flags. How many ways can all 11 flags be flown?

Answer

There will be one pole with 3 flags, and 4 flags on the other 2 poles. There are 3 ways to choose the 3-flag pole, and 11! orders to arrange the flags, giving $3 \cdot 11!$ ways to fly the flags.

2.2.7. How many ways can a two scoop ice cream cone be ordered from 10 flavors if

(a) the two flavors are different and it matters which flavor is on top?

(b) the two flavors are different and it doesn't matter which flavor is on top (e.g., chocolate over strawberry is the same as strawberry over chocolate)?

(c) both scoops can be the same flavor but it still matters which flavor is on top?

(d) both scoops can be the same flavor, but if two different flavors are ordered it makes no difference which flavor is on top?

Answer

(a) $P(10, 2) = 10 \cdot 9 = 90$

(b) $C(10, 2) = 45$

(c) $P(10, 2) + 10 = 10 \cdot 9 + 10 = 100$

(d) $C(10, 2) + 10 = 45 + 10 = 55$

2.2.9. Every week, Jane rearranges the six pictures in her den, using the same hooks on the walls. What is the maximum number of weeks that can pass until some arrangement is repeated?

Answer

There are $6! = 720$ permutations, so some arrangement has been repeated at least after 721 weeks by the pigeonhole principle. This means Jane might have a new arrangement for nearly 14 years with careful planning!

2.2.11. Give a combinatorial proof of the identity $n^2 = 2\binom{n}{2} + n$ by asking a question that can be answered in two ways.

Answer

Sample question: "How many ways can two red checkers be placed on a $2 \times n$ checker board with one checker in each row?"

Answer 1. There are n choices for the placement of each checker, giving n^2 ways to position the two checkers.

Answer 2. There are two cases. If the checkers are in different columns, there are $\binom{n}{2}$ ways to choose the two columns and two ways—top row or bottom row—to place the checker in the left column. If both checkers are in the same column, there are n ways to choose the column. Altogether, there are $\binom{n}{2} 2 + n$ placements of the two checkers.

2.2.13. The nth Catalan number (named after Belgian mathematician Eugène Charles Catalan, 1814–1894; also see the next section for its meaning in the block walking model) is given by $C_n = \frac{1}{n+1}\binom{2n}{n}$. Prove that another formula is $C_n = \frac{1}{n}\binom{2n}{n-1}$, using both

(a) combinatorial reasoning

(b) algebraic calculation

Answer

(a) The two formulas for C_n are equal if it can be shown that $n\binom{2n}{n} = (n+1)\binom{2n}{n-1}$. This is true by using Theorem 2.8. A direct argument is given by choosing a committee of size n and its chair from a group of $2n$ people. Choosing the full committee and then its chair gives the left side, choosing all but the chair followed by a chair for the committee gives the right side.

(b)

$$C_n = \frac{1}{n+1}\binom{2n}{n} = \frac{(2n)!}{(n+1)n!n!}$$

$$= \frac{(2n)!}{(n+1)!n(n-1)!} = \frac{1}{n}\frac{(2n)!}{(n+1)!(n-1)!} = \frac{1}{n}\binom{2n}{n-1}$$

2.2.15. The United States Post Office automates mail handling with the Postnet Bar Code (postal numeric encoding technique). There are 62 vertical bars of two sizes, long and short. Between the two long guard bars at the beginning and end, there are 12 five-symbol blocks that each contain two long and three shorts bars:

7 4 8 0 5 3 7 5 8 1 3 9

For example, the first block of five symbols that follows the leftmost guard bar is long-short-short-short-long, which encodes the digit 7.

(a) Explain why a blocks of two long and five short symbols works well.

(b) What blocks do you expect code the digits 2 and 6?

Answer

(a) $\binom{5}{2} = \frac{5 \cdot 4}{2} = 10$, so there are 10 different blocks, just right to code the 10 digits $1, 2, 3, \ldots, 9, 0$.

(b) The missing permutations are shown below.

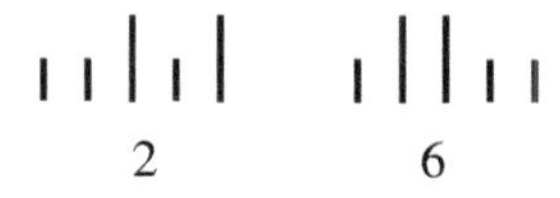

2.2.17. Place n points on a circle and then construct all of the chords that join pairs of points. Assume your n points are in "general position," so no more than two chords intersect at a single point within the circle. The following diagram depicts the case $n = 7$, and shows that there are 21 chords and 35 intersections of chords within the circle:

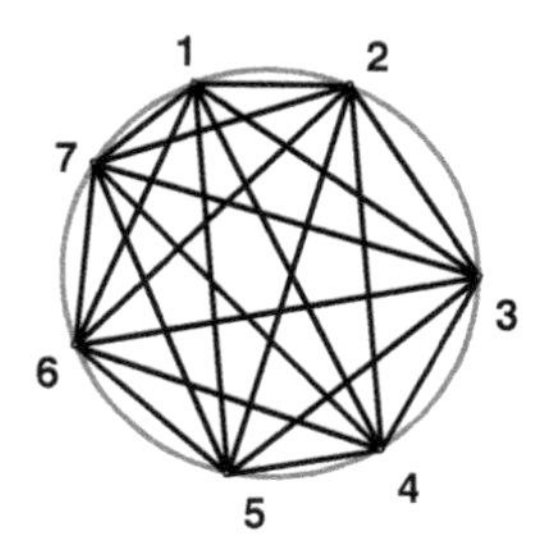

For n points, what is

(a) the number of chords?

(b) the number of points of intersection of chords within the circle?

Answer

(a) Every pair of points determines a chord, so there are $\binom{n}{2}$ chords.

(b) Every crossing point corresponds to 4 of the n points, and conversely, so there are $\binom{n}{4}$ points at which chords intersect within the circle.

2.2.19. How many paths connect A to B in the system of blocks shown (see diagram) that surround the lake in the center?

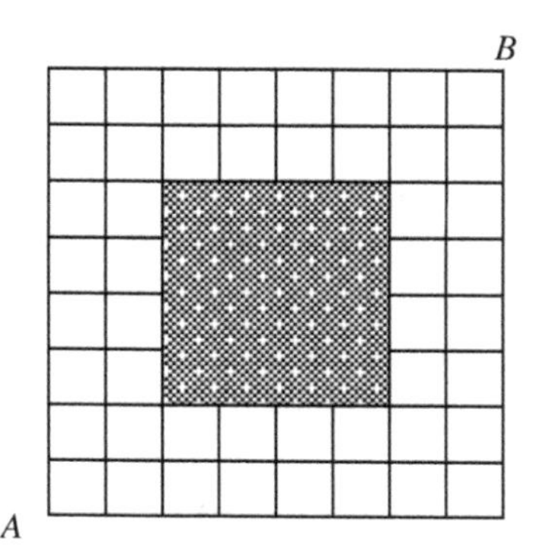

Answer

Any path passing below the lake must pass through exactly one of the points (6, 2), (7, 1), or (8, 0). There are $\binom{8}{2}$ paths from $A(0, 0)$ to (6, 2), and similarly the same number of paths from (6, 2) to $B(8, 8)$. Therefore, there are $\binom{8}{2}^2$ such paths from A to B passing through (6,2). Similar reasoning applies to the paths through (7, 1) and (8, 0), as well as the paths through (2, 6), (1, 7), and (0, 8) that pass above the lake. Altogether, there are $\binom{8}{0}^2 + \binom{8}{1}^2 + \binom{8}{2}^2 + \binom{8}{6}^2 + \binom{8}{7}^2 + \binom{8}{8}^2$ paths from A to B that skirt the lake.

2.2.21. A bridge hand consists of 13 cards selected from a 52 card deck. How many bridge hands

(a) are there?

(b) do not contain a pair? (That is, no two cards have the same rank.)

(c) contain *exactly* one pair?

Answer

(a) $\binom{52}{13}$

(b) Every rank must be present in the hand, and there are $\binom{4}{1} = 4$ ways to choose one card of any rank. Therefore, 4^{13} bridge hands have no pair.

(c) The rank of the pair can be chosen in 13 ways, and a pair of this rank in $\binom{4}{2}$ ways. The 11 remaining cards must be of all different ranks chosen from the 12 ranks different from the pair, and there are 4 choices of the suit from each rank. Altogether, there are $\binom{13}{1}\binom{4}{2}\binom{12}{11}4^{11}$ bridge hands with exactly one pair.

2.2.23. Consider the identity $(2n-1)!! = \frac{n!}{2^n}\binom{2n}{n}$, where $(2n-1)!! = (2n-1)\cdot(2n-3)\cdot\cdots\cdot 3\cdot 1$ is a "double" factorial.

(a) Prove the identity combinatorially, by counting the number of ways to set up singles tennis matches for $2n$ players.

(b) Prove the identity algebraically.

Answer

(a) Suppose the players are numbered 1 through $2n$. Player $2n$ can choose an opponent in $2n - 1$ ways. The next highest numbered player that remains can choose an opponent in $2n-3$ ways and so on. That is, there are $(2n-1)!!$ ways to set up the n pairs of opponents. Alternatively, in $\binom{2n}{n}$ ways, choose n of the $2n$ players to go out onto the n courts, one player per court. Their opponents can be assigned in $n!$ ways. The same matchups occur if the two players on any court were to trade places, so altogether there are $\dfrac{\binom{2n}{n} n!}{2^n}$ ways to set up the matches.

(b) $$\frac{n!}{2^n}\binom{2n}{n} = \frac{n!(2n)!}{2^n n! n!} = \frac{(2n)(2n-1)(2n-2)\cdots(4)(3)(2)(1)}{2^n n!}$$
$$= \frac{(n)(2n-1)(n-1)\cdots(2)(3)(1)(1)}{n!}$$
$$= (2n-1)(n-3)\cdots(3)(1) = (2n-1)!!$$

2.2.25. Would you want to bet for or against getting 9, 10 or 11 heads in 20 flips of a fair coin?

Answer

$Prob(9,\ 10,\ \text{or}\ 11\ \text{heads}) = \dfrac{\binom{20}{9}+\binom{20}{10}+\binom{20}{11}}{2^{20}} = 0.4966$, so it is just slightly better to bet against getting 9, 10, or 11 heads.

PROBLEM SET 2.3

2.3.1. Recall that the combinatorial Fibonacci number f_n is the number of ways to tile a $1 \times n$ board with squares and dominoes. Define $f_0 = 1$ and use the tiling model to prove the identity $f_{2n+1} = f_0 + f_2 + f_4 + \cdots + f_{2n}$, $n \geq 0$.

Answer

The left side f_{2n+1} counts the number of ways to tile a board of odd length, which requires at least one square tile. Now take cases that depend on the position of the right most square, and suppose that k dominoes are to its right. To the left of the rightmost square there is a board of length $(2n+1) - (1+2k) = 2(n-k)$ that can be tiled in $f_{2(n-k)}$ ways. Summing over $k = 0, 1, 2, \ldots, n$ gives the right side of the identity.

2.3.3. **(a)** Prove that $f_{2n+1} = \sum_{k=1}^{n+1} \binom{n+1}{k} f_{k-1}$, $n \geq 0$, using the tiling model.

[*Hint*: Any tiling of a $1 \times (2n+1)$ board must include at least one square and at least n other tiles.]

(b) Restate the identity of part (a) in terms of the standard Fibonacci numbers $F_m = f_{m-1}$.

Answer

(a) The left side f_{2n+1} counts the number of ways to tile a board of length $2n+1$. Notice that any tiling includes at least one square and n other tiles. Let k be the number of squares in the first $n + 1$ tiles of the board, whose positions can be chosen in $\binom{n+1}{k}$ ways. The remaining $n + 1 - k$ tiles are dominoes, so the first $n + 1$ tiles cover a board of length $k + 2(n + 1 - k) = 2n + 2 - k$. This leaves a board of length $(2n + 1) - (2n + 2 - k) = k - 1$ that can be tiled in f_{k-1} ways. Therefore, there are $\binom{n+1}{k} f_{k-1}$ tilings of the board of length $2n + 1$ with k squares among the first $n + 1$ tiles. Summing over k gives the right side of the identity.

(b) $F_{2n+2} = \sum_{k=1}^{n+1} \binom{n+1}{k} F_k$, or more simply, $F_{2n} = \sum_{k=1}^{n} \binom{n}{k} F_k$.

2.3.5. Use the tiling model to prove that $F_{n+1}^2 - F_{n-1}^2 = F_{2n}$.

[*Hint*: Consider tilings of two boards each of length n, not both ending with a domino.]

Answer

First, rewrite the identity in terms of the combinatorial Fibonacci numbers, where it takes the form. $f_n^2 - f_{n-2}^2 = f_{2n-1}$. Then f_n^2 is the number of ways to tile two boards each of length n. Of these, f_{n-2}^2 tilings each end with a domino, so $f_n^2 - f_{n-2}^2$ counts the number of tilings of the two boards for which at least one ends in a square. If the first board ends with a square, it can be deleted to leave a board of length $n - 1$ and placed at the front of the second board to give a tiling of a board of length $2n - 1$ with a break following square $n - 1$. If the first board ends with a domino, then the second board ends with a square. Now joining the first board and the second board and deleting the rightmost square leaves a tiling of the board of length $2n - 1$ with no break between square $n - 1$ and square n. All of the boards of length $2n - 1$ are formed exactly once, so there are f_{2n-1} tilings.

2.3.7. Prove the identity below by providing two ways to answer this question: "In how many way can a $2 \times n$ board be tiled with n red and n blue square tiles?"

$$\binom{2n}{n} = \sum_{k \geq 0} \binom{n}{k} \binom{n-k}{k} 2^{n-2k}$$

Answer

Answer 1. There are $\binom{2n}{n}$ ways to place the n red tiles, with the remaining squares covered by blue tiles. This is the left side of the identity.

Answer 2. There are four types of columns: *rr, bb, rb,* and *br,* where the colors of the upper and lower tiles are listed in order. Suppose there are k all red columns of type *rr.* Since there are an equal number of red and blue tiles, there must also be k all blue columns. The remaining $n-2k$ columns can each be tiled with a red over a blue tile, *rb,* or with a blue over a red tile, *br.* The all red columns can be selected in $\binom{n}{k}$ ways, the all blue columns in $\binom{n-k}{k}$ ways, and the remaining $n-2k$ columns can be tiled in 2^{n-2k} ways. Summing over k gives the right side of the identity.

2.3.9. Use the block walking model to prove that $\binom{2n}{n} = \sum_{k=0}^{n} \binom{n}{k}^2$.

Answer

There are $\binom{2n}{n}$ paths from $A(0, 0)$ to $B(2n, n)$. Partition these paths by the unique point at diagonal k at which the path crosses row n. There are $\binom{n}{k}$ ways to reach this point from $A(0, 0)$, and the same number of paths continuing from this point to $B(2n, n)$. That is, there are $\binom{n}{k}^2$ paths from A to B that cross row n at diagonal k. The identity then follows by summing over all k.

2.3.11. There have been 100 East-versus-West bowl games, with 50 wins for each. What is the probability that West never had fewer wins than East throughout the series?

Answer

The bowl series can be modeled with block walking through Pascal's Triangle, so there are $\binom{100}{50}$ paths that represent the sequence of wins and losses. The Catalan number $C_{50} = \frac{1}{50+1}\binom{100}{50}$ is the number of paths for which West never trailed East in games won. Assuming that every path is equally likely, the probability of this is 1/51.

2.3.13. Give a block walking proof that the Catalan numbers described in Example 2.17 satisfy the recursion relation $C_{n+1} = C_0C_n + C_1C_{n-1} + \cdots + C_{n-1}C_1 + C_nC_0$, where $C_0 = 1$.

Answer

Let $B = B(2n + 2, n + 1)$ be the point in row $2n + 2$ crossed by diagonal $n + 1$. The Catalan number C_{n+1} counts the paths from $A(0, 0)$ to B

that never cross the vertical line through A and B. However, that path can return to the line, say for the first time at the point $F(2r + 2, r + 1)$ where $0 \leq r \leq n$. This means the path from $P(1, 0)$ to $Q(2r + 1, r)$ never crosses the vertical line from P to Q, so there are C_r of these paths. There are also C_{n-r} paths from F to B. We see there are $C_r C_{n-r}$ paths from A to B that first return to the line AB at the point F in row r. Summing over r gives the desired sum.

2.3.15. Use the committee selection model to prove the identity $\binom{2n}{n} = \sum_{k=0}^{n} \binom{n}{k}^2$.

Answer

Answer 1. In a club of n men and n women, there are $\binom{2n}{n}$ ways to select a committee of n members with no attention to the gender of the committee members.

Answer 2. There are $\binom{n}{k}$ ways to select k men on the committee and then fill out the committee by selecting, in $\binom{n}{k}$ ways, k of the n women *not* to put on the committee. This means that there are $\binom{n}{k}^2$ ways to select an n-person committee with k men and $n - k$ women. Summing over k shows there are $\sum_{k=0}^{n} \binom{n}{k}^2$ ways to form the committee.

2.3.17. Prove that the binomial coefficients in row n of Pascal's triangle are *unimodal,* meaning that they strictly increase to a maximum $\binom{2m}{m}$ when $n = 2m$ is even and then strictly decrease, or, strictly increase to the maximum $\binom{2m-1}{m} = \binom{2m-1}{m+1}$ when $n = 2m - 1$ is odd, and then strictly decrease.

Answer

By the symmetry of Pascal's triangle, it suffices to show that $\binom{n}{r-1} < \binom{n}{r}$ for all r for which $1 \leq r \leq n/2$. This follows from $\dfrac{\binom{n}{r}}{\binom{n}{r-1}} = \dfrac{n+1-r}{r} > \dfrac{n-r}{r} \geq 1$.

2.3.19. Use the flagpole model to prove the identity, $\sum_{k=1}^{n} k \binom{n}{k} = n\, 2^{n-1}$, where $n \geq 1$, by giving two answers to this question: "In how many ways can a flagpole be placed on one of n blocks, with any subset (possibly empty) of the remaining blocks used to anchor one guy wire each?"

Answer

Answer 1. There are n choices of the block to support the pole, and 2^{n-1} subsets of the remaining $n - 1$ blocks on which to anchor guy wires. This is the right side of the identity.

Answer 2. There are $\binom{n}{k}$ ways to choose k blocks to be used in the arrangement, and the one supporting the flag can then be chosen in k ways. Summing over k gives the left side of the identity.

PROBLEM SET 2.4

2.4.1. How many words (i.e., words or nonword letter combinations) can be formed using all of the letters of the word combinatorics if

(a) there are no restrictions?

(b) the three-letter sequence bat must appear in the word?

(c) the two-letter sequence cc does not appear?

Answer

(a) $13!/(2!2!2!)$

(b) View bat as a single "letter", so there are $11!/(2!2!2!)$ "words."

(c) $\dfrac{13!}{2!2!2!} - \dfrac{12!}{2!2!}$

2.4.3. **(a)** How many paths join (0, 0, 0) to (3, 3, 3) where each step along the path has the form (1, 0, 0), (0, 1, 0) or (0, 0, 1)?

(b) How many paths that join (0, 0, 0) to (3, 3, 3) include the opposite corners (1, 1, 1) and (2, 2, 2) of the inner unit cube?

Answer

(a) $\binom{9}{3,3,3}$

(b) $\binom{3}{1,1,1}^3 = (3!)^3 = 6^3$

2.4.5. **(a)** Show, by numerical evaluations, that

$$\binom{12}{3,5,4} = \binom{11}{2,5,4} + \binom{11}{3,4,4} + \binom{11}{3,5,3}$$

(b) Explain why the result in part (a) holds by combinatorial reasoning.

(c) Prove the identity

$$\binom{n}{a, b, c, \ldots, z} = \binom{n-1}{a-1, b, c, \ldots, z} + \binom{n-1}{a, b-1, c, \ldots, z} + \cdots + \binom{n-1}{a, b, \ldots, z-1}$$

Answer

(a) $27{,}720 = 6930 + 11{,}550 + 9240$

(b) The last step to (3, 5, 4) must be from (2, 5, 4), or (3, 4, 4), or (3, 5, 3).

(c) The last step to $(a, b, c, \ldots, z)$ must be from $(a-1, b, c, \ldots, z)$ or $(a, b-1, c, \ldots, z) \ldots$, or $(a, b, c, \ldots, z-1)$.

2.4.7. A well-stocked bakery sells four kinds of donuts—maple, glazed, jelly, and chocolate. How many ways can you buy a dozen of the items

(a) with no restrictions?

(b) provided there must be at least two glazed and three jelly donuts?

(c) provided there are no more than two maple bars?

Answer

(a) $\left(\!\binom{4}{12}\!\right) = \binom{12+4-1}{12} = \binom{15}{3} = 455$

(b) $\left(\!\binom{4}{12-2-3}\!\right) = \binom{7+4-1}{7} = \binom{10}{3} = 120$

(c) Taking cases that depend on choosing 0, 1, or 2 maple bars, there are $\left(\!\binom{3}{12}\!\right) + \left(\!\binom{3}{11}\!\right) + \left(\!\binom{3}{10}\!\right) = \binom{14}{2} + \binom{13}{2} + \binom{12}{2} = 235$ ways to place the order. Alternatively, subtracting orders with 3 or more maple bars gives $\left(\!\binom{4}{12}\!\right) - \left(\!\binom{4}{12-3}\!\right) = 455 - \binom{12}{9} = 455 - 220 = 235$.

2.4.9. Find the number of integer solutions of the equation

$$x_1 + x_2 + x_3 + x_4 = 15, x_1 \geq 2, x_2 \geq 0, x_3 \geq 0, x_4 \geq 0$$

Answer

Let $x_1 = y_1 + 2$. Then $y_1 + x_2 + x_3 + x_4 = 13, y_1 \geq 0, x_2 \geq 0, x_3 \geq 0, x_4 \geq 0$. Hence there are $\left(\!\binom{4}{13}\!\right) = \binom{16}{13} = 560$ solutions.

2.4.11. Find the number of integer solutions of the equation
$e_1 + e_2 + e_3 + e_4 = n, 2 \le e_1 \le 5, 0 \le e_2, 0 \le e_3, 0 \le e_4$.

Answer

Let $e_1 = e_1' + 2$ to obtain the equation $e_1' + e_2 + e_3 + e_4 = n - 2$, $0 \le e_1' \le 3, 0 \le e_2, 0 \le e_3, 0 \le e_4$. Therefore, there are $\left(\!\binom{4}{n-2}\!\right) - \left(\!\binom{4}{n-2-4}\!\right) = \binom{n+1}{3} - \binom{n-3}{3}$ solutions.

2.4.13. Apple and berry pies each call for a cup of sugar, but a rhubarb pie requires 3 cups of sugar. How many ways can pies be made if all 10 cups of sugar that are available must be used?

Answer

We have the equation $x_a + x_b + 3x_r = 10$ in nonnegative variables that specify the number of pies of each type. Clearly $x_r \in \{0, 1, 2, 3\}$, giving the four equations $x_a + x_b = 10, x_a + x_b = 7, x_a + x_b = 4, x_a + x_b = 1$ which have altogether $11 + 8 + 5 + 2 = 26$ solutions.

2.4.15. How many permutations of the multiset $S = \{3 \cdot \mathsf{A}, 5 \cdot \mathsf{B}, 2 \cdot \mathsf{C}, 2 \cdot \mathsf{D}\}$ have no adjacent As?

Answer

The permutations have the form $\boxed{x_1}\,\mathrm{A}\,\boxed{x_2}\,\mathrm{A}\,\boxed{x_3}\,\mathrm{A}\,\boxed{x_4}$, where x_i is the number of letters preceding, between, or following the As. Thus, $x_1 + x_2 + x_3 + x_4 = 9$, $x_1, x_4 \ge 0, x_2, x_3 \ge 1$, which has $\left(\!\binom{4}{9-1-1}\!\right) = \binom{10}{7}$ solutions. The letters in $\{5 \cdot \mathsf{B}, 2 \cdot \mathsf{C}, 2 \cdot \mathsf{D}\}$ have $\binom{9}{5,2,2}$ permutations, so the number of permutations of S with no adjacent As is $\binom{10}{7}\binom{9}{5,2,2}$.

2.4.17. Let u_n denote the number of binary sequences of As and Bs of length n with no adjacent As. For example, there are $u_2 = 3$ sequences of length 3, namely AB, BA, and BB.

(a) How many of these sequences have precisely k As? Use the observation that any A must be followed with a B unless it is at the end of the sequence.

(b) Solve part (a) by counting the number of integer solutions of an equation.

(c) Derive a recursion formula for the u_n sequence.

(d) Identify the well-known sequence given by the recursion.

(e) What identity that follows from parts (a) and (d)?

Answer

(a) If the binary sequence ends with a B then the k As must always occur as the pair AB. This means the sequence is a permutation of k AB pairs and $n - 2k$ Bs. There are $k + n - 2k = n - k$ symbols and therefore $\binom{n-k}{k}$ ways to place the AB pairs. If the sequence ends with an A, then the sequence is a permutation of $k - 1$ AB pairs and $n - 2(k - 1) - 1 = n - 2k + 1$ Bs followed by the A at the end. There are $\binom{n-2k+1+k-1}{k-1} = \binom{n-k}{k-1}$ such sequences. The total number of sequences with exactly k As, no two adjacent, is therefore $\binom{n-k}{k} + \binom{n-k}{k-1} = \binom{n-k+1}{k}$ by Pascal's identity.

(b) Let x_i be the number of Bs that precede the ith A, and let x_{k+1} be the number of Bs following the last A. Then

$$x_1 + x_2 + \cdots + x_{k+1} = n - k, \quad x_1, x_{k+1} \geq 0, \quad x_2, x_3, \ldots, x_k \geq 1.$$

There are $\left(\!\!\binom{k+1}{(n-k)-(k-1)}\!\!\right) = \binom{n-k+1}{n-2k+1} = \binom{n-k+1}{k}$ solutions of the equation.

(c) If a sequence of length n begins with an A, the next symbol is a B and the remaining part of the sequence is any of the u_{n-2} sequences of length $n - 2$. If the sequence of length n begins with a B, the remaining symbols form any of the u_{n-1} sequences of length $n - 1$. Thus, $u_n = u_{n-2} + u_{n-1}$.

(d) The recursion derived in part (c) is the same as enjoyed by the Fibonacci numbers. However, $u_1 = 2 = F_3$ and $u_2 = 3 = F_4$, so we see that $u_n = F_{n+2}$.

(e) The number of sequences of As and Bs of length n with no two consecutive As is given both by the Fibonacci number F_{n+2} and by the sum $\sum_{k\geq 0} \binom{n-k+1}{k}$. Replacing n with $n - 1$ we arrive at the identity $F_{n+1} = \sum_{k\geq 0} \binom{n-k}{k}$.

2.4.19. The *Delannoy numbers* (see Problem 2.3.12) of the form $D(n, n)$ are also known as the "king's walk" numbers, since they give the number of paths that the king in chess can move from one corner to the opposite corner of a square chess board. Prove that

$$D(n,n) = \sum_{k\geq 0} \binom{n+k}{k}\binom{n}{k}.$$

Answer

The left side of the identity counts the number of paths from A to $B(n, n)$. Next, consider those paths with k southwest edges. There are also k southeast edges and $n - k$ south edges, so there are $k + k + n - k = n + k$ edges in all. Thus there are $\binom{n+k}{k, k, n-k} = \binom{n+k}{k}\binom{n}{k}\binom{n-k}{n-k} = \binom{n+k}{k}\binom{n}{k}$ paths from A to B with k southwest edges. Summing over k gives the right side of the identity.

2.4.21. Show that the number of permutations of at most m As and at most n Bs is $\binom{m+n+2}{m+1} - 2$, where $m \geq 0$, $n \geq 0$ and $m + n > 0$.

Answer

Given any permutation of $m + 1$ As and $n + 1$ Bs, delete the leftmost B and any As to its left together and similarly delete the rightmost A together with any Bs to its right. Of course, the permutation must have a B to the left of an A, so we must not start with the permutation AA ... ABB ... B for which all of the $m + 1$ As are to the left of the $n + 1$ Bs. This leaves a permutation of at most m As and at most n Bs. However, to have $m + n > 0$, the leftmost B cannot be adjacent to the rightmost A, so we cannot use the permutation AA ... ABAB ... B. Conversely, any permutation of at most m As and at most n Bs can be extended by appending a string AA ... AB at its right end and a string AB ... B at its left end. This gives a one-to-one correspondence with the $\binom{m+n+2}{m+1} - 2$ permutations of $m + 1$ As and $n + 1$ Bs other than AA ... ABB ... B and AA ... ABAB ... B.

2.4.23. Prove that $n\left(\!\binom{k}{n}\!\right) = k\left(\!\binom{k+1}{n-1}\!\right)$ [*Hint*: Replace the Δ in any permutation of the multiset $S = \{(n-1)\cdot *, (k-1)\cdot |, 1\cdot \Delta\}$ with either a star or a bar.]

Answer

Any permutation of $S = \{(n-1)\cdot *, (k-1)\cdot |, 1\cdot \Delta\}$ corresponds to one of the $n\left(\!\binom{k}{n}\!\right)$ permutations of $\{n\cdot *, (k-1)\cdot |\}$ in which one of the n stars is replaced with a Δ. Similarly, any permutation of the multiset S

corresponds to one the $k\left(\left(\binom{k+1}{n-1}\right)\right)$ permutations of $\{(n-1)\cdot *, k\cdot |\}$ in which one of the k bars is replaced with a Δ. The result also follows algebraically:

$$n\left(\left(\binom{k}{n}\right)\right) = n\binom{n+k-1}{n} = \frac{n(n+k-1)!}{n!(k-1)!} = \frac{k(n+k-1)!}{(n-1)!k!}$$

$$= k\binom{n-1+(k+1)-1}{n-1} = k\left(\left(\binom{k+1}{n-1}\right)\right)$$

2.4.25. **(a)** Use combinatorial reasoning to prove that

$$\left(\left(\binom{k+1}{n+1}\right)\right) = \sum_{r=0}^{k}\left(\left(\binom{r+1}{n}\right)\right)$$

[*Hint*: The left side of the identity counts the number of ways to order $n+1$ ice cream cones chosen from $k+1$ flavors.]

(b) Identify the identity proved in part (a) when it is written in terms of binomial coefficients.

Answer

(a) Suppose that the order for ice cream includes at least one cone of flavor $r+1$, and no cone of a flavor numbered larger than $r+1$. There are then $\left(\left(\binom{r+1}{n}\right)\right)$ ways to order the remaining n cones from flavors $1, 2, \ldots, r+1$. Summing over r, $0 \le r \le k$, gives the right side of the identity.

(b) It is the Hockey Stick identity

$$\binom{n+k+1}{n+1} = \left(\left(\binom{k+1}{n+1}\right)\right) = \sum_{r=0}^{k}\left(\left(\binom{r+1}{n}\right)\right) = \sum_{r=0}^{k}\binom{n+r}{n}.$$

PROBLEM SET 2.5

2.5.1. How many ways can 15 identical candies be given to 3 children so each receives an odd number of candies?

Answer

First, give each child one candy. The remaining 12 candies which can be distributed as 6 pairs in $\left(\left(\binom{3}{6}\right)\right) = \binom{6+3-1}{6} = \binom{8}{6} = \frac{8\cdot 7}{2!} = 28$ ways.

2.5.3. **(a)** In how many ways can six identical balls be distributed to 10 people?

(b) Propose a selection problem at the ice cream store that is equivalent to the problem of part (a).

(c) What equation (in integers) is equivalent to the problems in parts (a) and (b)?

Answer

(a) $\left(\!\!\binom{10}{6}\!\!\right) = \binom{6+10-1}{6} = \binom{15}{6}$

(b) In how many ways can an order be placed for 6 ice cream cones at a store offering 10 flavors.

(c) $x_1 + x_2 + \cdots + x_{10} = 6, x_i \geq 0.$

2.5.5. Use Table 2.1 and the triangle identity (2.23) to calculate the distribution number $T(8, 4)$.

Answer

$$\begin{aligned} T(8,4) &= 4T(7,3) + 4T(7,4) \\ &= 4(3T(6,2) + 3T(6,3)) + 4(4T(6,3) + 4T(6,4)) \\ &= 12T(6,2) + 28T(6,3) + 16T(6,4) \\ &= 12 \cdot 62 + 28 \cdot 540 + 16 \cdot 1560 = 40,824 \end{aligned}$$

2.5.7. For each expression listed below, discuss whether or not it answers the question "In how many ways can 4 different books be placed on a bookcase with three shelves, where no shelf remains empty?" Give the reasoning that led to the answer, and explain why it is correct or incorrect.

(a) $T(4,3) \cdot 4!$

(b) $T(4,3) \cdot 2!$

(c) $\left(\!\!\binom{3}{1}\!\!\right) \cdot 4!$

Answer

(a) "Distribute the books in $T(4, 3)$ ways that leave no shelf empty, then permute the books in 4! ways." This overcounts the arrangements. For example, both distributions $\boxed{AB}\boxed{C}\boxed{D}$ and $\boxed{AB}\boxed{D}\boxed{C}$ have permutations that give the same arrangement $AB|C|D$.

(b) "Distribute the books in $T(4, 3)$ ways that leave no shelf empty, so there are 2 shelves with a single book and two books on a third shelf that can be placed in 2! orders." This is a correct answer, and shows there are 72 arrangements.

(c) "Let x_i be the number of books on shelf i, so $x_1 + x_2 + x_3 = 4, x_i \geq 1$. This equation has $\left(\!\!\binom{3}{4-1-1-1}\!\!\right) = \left(\!\!\binom{3}{1}\!\!\right) = 3$ solutions. Now place the books in 4! ways on the shelves. Again there are $3 \cdot 4! = 72$ arrangements." This is also correct reasoning.

2.5.9. Use combinatorial reasoning to prove that $T(m, 2) = 2^m - 2$.

Answer

There are 2^m ways to distribute m distinct objects to two distinct recipients, including the 2 distributions in which one recipient is assigned all m objects. By the subtraction principle, there are $2^m - 2$ onto distributions.

2.5.11. Use combinatorial reasoning to prove that

$$T(n+3, n) = n!\left[\binom{n+3}{4} + 10\binom{n+3}{5} + 5 \cdot 3\binom{n+3}{6}\right]$$

Answer

Distribute $n + 3$ distinct objects to n people so that every person receives at least one object.

There are three cases to consider:

Case 1. One recipient receives four objects

There are $\binom{n+3}{4}$ ways to bundle 4 of the $n + 3$ objects and then $n!$ ways to distribute the bundle of 4 and the $n - 1$ single objects to the n recipients.

Case 2. One recipient receives 3 objects and another receives 2 objects

There are $\binom{n+3}{5}$ ways to choose 5 objects that can then be bundled as a pair and a triple in $\binom{5}{2} = 10$ ways. The two bundles and the $n - 2$ single objects can then be distributed in $n!$ ways.

Case 3. Three recipients each receive a pair of objects

There are $\binom{n+3}{6}$ ways to choose 6 objects that can be divided into three pairs in $5 \cdot 3$ ways. The three pairs and the $n - 3$ single objects can then be distributed in $n!$ ways.

2.5.13. How many permutations of the letters I, N, F, I, N, I, T, I, E, S do not contain any adjacent pair of Is?

Answer

The four letters I must be separated by other letters, so the permutation must have the form $\boxed{x_1}\,\mathrm{I}\,\boxed{x_2}\,\mathrm{I}\,\boxed{x_3}\,\mathrm{I}\,\boxed{x_4}\,\mathrm{I}\,\boxed{x_5}$, where $x_1 \geq 0, x_2 \geq 1, x_3 \geq 1, x_4 \geq 1, x_5 \geq 0$ denote number of letters in the corresponding positions of the permutation. That is, we first count the number of solutions in integers of the equation $x_1 + x_2 + x_3 + x_4 + x_5 = 6, x_1 \geq 0, x_2 \geq 1, x_3 \geq 1, x_4 \geq 1, x_5 \geq 0$. This number is $\left(\!\binom{6}{6-1-1-1}\!\right) = \left(\!\binom{6}{3}\!\right) =$

$\binom{3+6-1}{3} = \binom{8}{3} = 56$. Also, there are $6!/2 = 360$ permutations of the multiset $\{\mathsf{F}, 2 \cdot \mathsf{N}, \mathsf{T}, \mathsf{E}, \mathsf{S}\}$. Altogether, this gives $56 \cdot 360 = 20{,}160$ permutations of the 10 letters in which no two Is are adjacent.

2.5.15. **(a)** Given any polynomial $p(x) = a_0 + a_1x + a_2x^2 + \cdots + a_nx^n$, prove there are constants $b_0, b_1, \ldots, b_n$ so that $p(k) = b_0\binom{k}{0} + b_1\binom{k}{1} + b_2\binom{k}{2} + \cdots + b_n\binom{k}{n}$ for all positive integers k.

(b) Find the constants b_0, b_1, b_3, b_3 for which $k^3 = b_0\binom{k}{0} + b_1\binom{k}{1} + b_2\binom{k}{2} + b_3\binom{k}{3}$.

[*Note*: It can be shown (see Problem 2.5.18) that $b_r = \sum_{k=r}^{n} T(k, r)\, a_k$.]

Answer

(a) Proof by induction on n. For $n = 0$, $p(0) = a_0 = a_0\binom{0}{0}$, so $a_0 = b_0$. Now assume every polynomial of degree less than n can be written in the required form. Then let $b_n = n!a_n$, so $p(k) - b_n\binom{k}{n} = a_0 + a_1k + a_2k^2 + \cdots + a_nk^n - a_nk(k-1)(k-2)\cdots(k-n+1)$ is a polynomial of degree less that n. By the induction hypothesis, there are constants $b_0, b_1, \ldots, b_{n-1}$ so that $p(k) - b_n\binom{k}{n} = b_0\binom{k}{0} + b_1\binom{k}{1} + \cdots + b_{n-1}\binom{k}{n-1}$, and we see that $p(k)$ has the form required.

(b) It is known from part (a) that $k^3 = 0\binom{k}{0} + b_1\binom{k}{1} + b_2\binom{k}{2} + 3! \cdot 1\binom{k}{3} = b_1\binom{k}{1} + b_2\binom{k}{2} + 6\binom{k}{3}$. For $k = 1$, we get $1 = b_1$ and for $k = 2$, we get $2^3 = 2b_1 + b_2$ so $b_2 = 8 - 2 = 6$.

Therefore, $k^3 = \binom{k}{1} + 6\binom{k}{2} + 6\binom{k}{3}$.

2.5.17. Use the result of Problem 2.5.16 to prove the polynomial identity

$$x^m = \sum_{n=0}^{m} T(m, n)\,\frac{(x)_n}{n!},$$

where m is a nonnegative integer, x is the variable of the polynomials, and $(x)_r = x(x-1)\cdots(x-r+1)$.

Answer

When $x = k$, any positive integer, we have $\frac{(k)_n}{n!} = \binom{k}{n}$. Therefore, by Problem 2.5.16, the polynomial $p(x) = x^m - \sum_{n=0}^{m} T(m,n)\frac{(x)_n}{n!} = 0$ when x is any positive integer. But the only polynomial with infinitely many zeroes is the 0 polynomial, so $p(x) = 0$ which proves the identity.

PROBLEM SET 2.6

2.6.1. There are eight people at a dinner party, including four men and four women. In how many ways can they be seated at a circular table if

(a) there are no restrictions?

(b) no two men can be seated side by side?

(c) Mr. and Mrs. Smith insist on sitting next to each other?

(d) George and Alice insist on not sitting next to one another?

Answer

(a) 7!

(b) The men can be seated in 3! ways, and then the women can be seated between the men in 4! ways, giving $3! \cdot 4!$ seatings.

(c) Seat all but Mr. Smith in 6! ways, and then seat Mr. Smith to either the left or right of Mrs. Smith in 2 ways, giving $6! \cdot 2$ seatings.

(d) Seat everyone but Alice in 6! ways, and then seat Alice in any one of the 5 seats not adjacent to George. Therefore, there are $6! \cdot 5$ seatings.

2.6.3. There are five adults and eight children at a party. In how many ways can they be seated at a circular table if at least one child is seated between any two adults?

Answer

The adults can be seated in 4! ways, leaving 5 distinct spaces between the adjacently seated adults. Let $x_1, x_2, \dots, x_5$ be the number of children to be seated between the adults, so that $x_1 + x_2 + x_3 + x_4 + x_5 = 8, x_i \geq 1$. The number of solutions (think "stars and bars") is then $\binom{8-5+5-1}{3} = \binom{7}{3} = 35$. Now line up the 8 children in 8! ways, and in clockwise order, seat the first x_1 children, then the next x_2 children, and so on to complete the seating. We now see there are $4! \cdot 35 \cdot 8!$ seatings.

2.6.5. **(a)** In how many ways can eight people be seated at a square table, with two persons on each side of the table?

(b) In how many ways can 12 people be seated at a square table, with three persons on each side of the table?

Answer

(a) First seat person #1 in 2 ways, either at the left or right of any of the identical sides. The other 7 people can be seated around the table in $7!$ ways, so there are $2 \cdot 7!$ seatings.

(b) There are now 3 ways—the left, middle, or right seat—to seat person #1, and $11!$ ways to seat the remaining people. This gives $3 \cdot 11!$ seatings.

2.6.7. How many permutations $\pi\colon [n] \to [n]$ satisfy $\pi(1) \neq 1$?

Answer

There are $n-1$ choices for $\pi(1)$ and $(n-1)!$ ways to permute the remaining $n-1$ elements. Thus, $(n-1)(n-1)!$ permutations do not fix element 1.

2.6.9. Players A and B each shuffle their own deck of 9 cards numbered 1, … , 9. Each player then shows a card. If the two cards match, player A wins, but if the two cards do not match then play continues with each player revealing another card. Player A wins the game if two matching cards are revealed on any turn, and player B wins the game if no match occurs after all 9 turns have been taken.

(a) Would you prefer to be player A or B?

(b) Would one of the players benefit if either 8 or 10 cards were used?

Answer

(a) Player B wins if his or her cards are a derangement of the sequence played by A, so the probability that B wins is $\frac{D_9}{9!} \approx \frac{1}{e} = 0.368$. Player A is far more likely to win, with a probability of about 0.632.

(b) By identity (2.29), $D_{10} = 10D_9 + 1$, so that $\frac{D_{10}}{10!} = \frac{D_9}{9!} + \frac{1}{10!} > \frac{D_9}{9!}$. Similarly, $\frac{D_9}{9!} = \frac{D_8}{8!} - \frac{1}{9!} < \frac{D_8}{8!}$. Therefore, player's B probability increases with either an 8 or 10 deck game, but the difference will scarcely be noticeable.

2.6.11. **(a)** How many permutations $\pi\colon [n] \to [n]$ derange k elements and leave the remaining elements fixed?

(b) Use combinatorial reasoning to prove the identity $n! = \sum_{k=0}^{n} \binom{n}{k} D_k$.

Answer

(**a**) Choose k elements in $\binom{n}{k}$ ways, and derange them in D_k ways. Therefore, there are $\binom{n}{k} D_k$ permutations of $[n]$ with $n - k$ fixed points and k deranged points.

(**b**) The $n!$ permutations of $[n]$ can be partitioned into disjoint subsets determined by the number, k, of elements that are deranged by the partition, where $0 \leq k \leq n$. This gives the identity by part (a) and the addition principle.

2.6.13. Use the identity of Problem 2.6.11(b) to show that $1 = \frac{1}{n!}\sum_{k=0}^{n} k\binom{n}{k} D_{n-k}$ for all n. That is, the mean μ (average) number of fixed points of the $n!$ permutations of $[n]$ is 1 (see Problem 2.6.12).

Answer

There are $\binom{n}{k} D_{n-k}$ permutations that have k fixed points. Thus, the average number of fixed points taken over all $n!$ permutations is

$$\begin{aligned}
\frac{1}{n!}\sum_{k=1}^{n} k\binom{n}{k} D_{n-k} &= \frac{1}{n!}\sum_{k=1}^{n} k\frac{n!}{k!(n-k)!}D_{n-k} = \sum_{k=1}^{n} \frac{1}{(k-1)!(n-k)!}D_{n-k} \\
&= \sum_{j=0}^{n-1} \frac{1}{(n-j-1)!j!}D_j = \frac{1}{(n-1)!}\sum_{j=0}^{n-1}\binom{n-1}{j}D_j \\
&= \frac{1}{(n-1)!}(n-1)! = 1.
\end{aligned}$$

In the calculation, there is a change of summation variable $j = n - k$, and the last summation is evaluated by the identity of Problem 2.5.12(b).

2.6.15. One morning, n monkeys lined up so that monkey 2 scratches the back of monkey 1 at the front of the line, monkey 3 scratches the back of monkey 2, and so on. That afternoon, the n monkeys want to line up again, but the new line up is such that no monkey's back will be scratched by the same monkey as in the morning. Let a_n denote the number of realignments of the n monkeys. For example, $a_1 = 1$ and $a_2 = 1$.

(**a**) Compile lists of the allowed realignments to determine that $a_3 = 3$ and $a_4 = 11$. For example, 3142 and 4231 are two allowable realignments if 1234 is the original order.

(**b**) Explain why there are $(n - 1)a_{n-1}$ ways to line up n monkeys, even if monkey n were to decide to leave the line.

(c) Explain why there are $(n-2)a_{n-2}$ ways to line up n monkeys, where monkey n cannot leave the newly formed line without creating a forbidden realignment.

(d) Summing parts (b) and (c) shows that $a_n = (n-1)a_{n-1} + (n-2)a_{n-2}$. Compare this to identity (2.27) to show that $a_n = \dfrac{D_{n+1}}{n}$.

Answer

(a) The three realignments of 123 are 132, 213, and 321.
The 11 realignments of 1234 are 1324, 1432, 2143, 2431, 2413, 3142, 3214, 3241, 4132, 4213, 4321.

(b) There are a_{n-1} realignments of the first $n-1$ monkeys, and monkey n can join the line in any of $n-1$ positions since the only forbidden position is immediately behind monkey $n-1$. This gives $(n-1)a_{n-1}$ realignments from which n can drop out.

(c) Monkey n is critical to forming an allowable realignment precisely when positioned between a successive pair $(j, j+1)$, $j = 1, 2, \ldots, n-2$ from the first alignment. There are $n-2$ such successive pairs, and a_{n-2} realignments of $n-2$ monkeys that retain the pair $(j, j+1)$. This means there are $(n-2)a_{n-2}$ alignments for which the presence of monkey n is essential to separate some pair of monkeys of the form j followed by $j+1$.

(d) The two cases described in parts (b) and (c) include all of the realignments, so summing gives the recursion relation $a_n = (n-1)a_{n-1} + (n-2)a_{n-2}$. Now multiply the recursion relation by n to get $na_n = n(n-1)a_{n-1} + n(n-2)a_{n-2}$, so that setting $na_n = d_{n+1}$ gives the recursion relation $d_{n+1} = nd_n + nd_{n-1}$. This is the same recursion relation shown in (2.28) that is satisfied by the derangement numbers D_{n+1}. Moreover, $d_2 = 1 \cdot a_1 = 1 = D_2$ and $d_3 = 2 \cdot a_2 = 2 \cdot 1 = 2 = D_3$. Starting with the same initial values and using the same recursion formula, the same sequence must be formed. That is, $na_n = D_{n+1}$ for all n.

PROBLEM SET 2.7

2.7.1. The factorials $n!$ grow very rapidly with n. Indeed, $15! = 1{,}307{,}674{,}368{,}000$ is already more than a trillion. In 1730, Abraham DeMoivre (1667–1754) gave the approximate value

$$n! \approx \left(\frac{n}{e}\right)^n \sqrt{2\pi n}$$

that is more often called *Stirling's formula* (James Stirling 1692–1770).

(a) Use Stirling's formula with a calculator to obtain an estimate of 15!.

(b) Show that a better approximation to 15! is given by Gosper's formula

$$n! \approx \left(\frac{n}{e}\right)^n \sqrt{\left(2n+\frac{1}{3}\right)\pi}$$

Answer

(a) Stirling's formula gives 1.30043×10^{12}.

(b) Gosper's formula gives 1.30763×10^{12}. It could have been predicted without a numerical evaluation that $15! = 1{,}307{,}674{,}368{,}000$ will end in 3 zeros. This is because 5 divides 5, 10, and 15 and no other factor of $15! = 15 \times 14 \times \cdots \times 2 \times 1$ and 2^3 is also a divisor of 15!, so $2^3 \cdot 5^3 = 10^3 = 1000$ divides 15!.

2.7.3. The nth *Catalan* number is given by $C_n = \frac{1}{n+1}\binom{2n}{n}$. Use Stirling's formula from Problem 2.7.1 to derive the approximate formula $C_n \approx \frac{4^n}{\sqrt{\pi}n^{3/2}}$.

Answer

$$C_n = \frac{1}{n+1}\frac{(2n)!}{(n!)^2} \approx \frac{1}{n}\frac{\left(\frac{2n}{e}\right)^{2n}\sqrt{4n\pi}}{\left(\left(\frac{n}{e}\right)^n\sqrt{2n\pi}\right)^2} = \frac{1}{n}4^n\frac{1}{\sqrt{\pi n}} = 4^n\frac{1}{\sqrt{\pi}n^{\frac{3}{2}}}$$

2.7.5. A *bridge hand* is a set of 13 cards dealt from a 52-card deck. How many bridge hands

(a) contain all four aces?

(b) contain four spades, four hearts, three diamonds, and two clubs?

(c) have a 5-3-3-2 distribution? That is, the hand contains five cards of one suit, three cards each of two other suits, and two cards from the remaining suit.

Answer

(a) $\binom{48}{9}$

(b) $\binom{13}{4}\binom{13}{4}\binom{13}{3}\binom{13}{2}$

(c) $\binom{4}{1}\binom{3}{1}\binom{13}{5}\binom{13}{3}^2\binom{13}{2}$. First choose which suit has 5 cards, then which suit has 2 cards, then select the cards from the suits.

2.7.7. **(a)** Explain why the number of paths from P_1 to P_n through the triangular grid shown, always moving in the direction of the arrows, is the nth Fibonacci number F_n:.

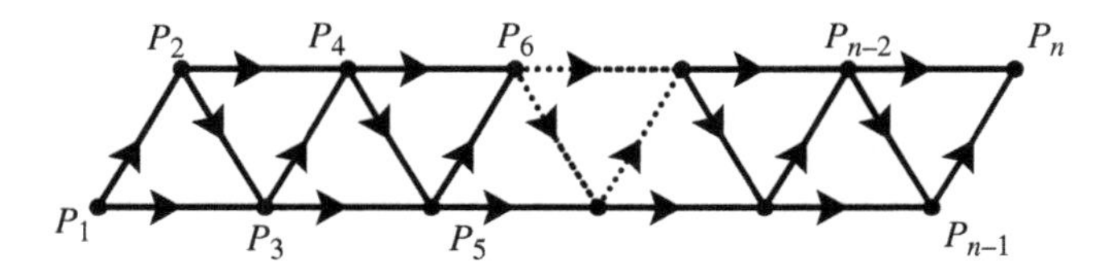

(b) How many paths connect point P_a to point P_b?

(c) Use the triangular grid to prove the identity $F_{a+b} = F_a F_{b+1} + F_{a-1} F_b$.

Answer

(a) There is one path to point P_1 (the null path), one path to point P_2, and 2 paths to point P_3, so we have the initial terms of the Fibonacci sequence. The number of paths that terminate at any point P_n is the sum of the number of paths that reach exactly one of the two penultimate points P_{n-1} or P_{n-2}, so we also have the Fibonacci recurrence $F_n = F_{n-1} + F_{n-2}$ for the number of paths. Thus the Fibonacci numbers give the number of paths to each point of the triangular grid.

(b) The number of paths is the same as the number of paths from P_1 to P_{b-a+1}, so there are F_{b-a+1} paths.

(c) By parts (a) and (b), there are F_{a+b} paths to point P_{a+b}, of which $F_a F_{b+1}$ paths pass through point P_a and $F_{a-1} F_b$ paths that avoid point P_a by traversing the edge $P_{a-1}P_{a+1}$.

2.7.9. Prove the identity

$$\binom{n}{p}\binom{n}{q} = \sum_{k\geq 0}\binom{n}{k}\binom{n-k}{p-k}\binom{n-p}{q-k}$$

by counting the number of ways to tile a $2 \times n$ board with p gray tiles in the top row, q gray tiles in the bottom row, and the remaining cells left empty. [*Hint*: Consider the number of columns that are tiled with two gray squares.]

Answer

Answer 1. The top row can be tiled in $\binom{n}{p}$ ways and the bottom row can be tiled in $\binom{n}{q}$ ways, so there are $\binom{n}{p}\binom{n}{q}$ ways to tile the board altogether.

Answer 2. Take cases depending on the number k of columns covered with 2 gray squares. There are $\binom{n}{k}$ ways to choose k of these columns, and then $\binom{n-k}{p-k}$ ways to choose the positions of the remaining $p-k$ gray tiles in the top row from the $n-k$ columns not completely tiled. Finally, there are $n-p$ positions for the additional $q-k$ gray tiles still needed for the bottom row, and these can be chosen in $\binom{n-p}{q-k}$ ways. Now sum over k.

2.7.11. Give a combinatorial proof that

$$\binom{n}{2}^2 = \binom{n}{2} + 6\binom{n}{3} + 6\binom{n}{4}$$

Answer

Various combinatorial models can be used. For example, "How many ways can 4 checkers be placed on a $2 \times n$ board, with two checkers in both the top and bottom row of squares?

Answer 1. There are $\binom{n}{2}$ ways to place the checkers in each row, giving $\binom{n}{2}^2$ ways to place the 4 checkers.

Answer 2. Take cases determined by the number of columns of the board that have at least one checker in the top or bottom row. If checkers appear in just two columns, there are $\binom{n}{2}$ ways to choose these columns and one way to place the 4 checkers. If there are checkers in three columns, there are $\binom{n}{3}$ ways to choose the columns, 3 ways to choose which column is covered with two checkers and 2 ways to put a checker into each of the two other columns. That is, there are $3 \cdot 2\binom{n}{3}$ arrangements with a single column covered with checkers. Finally, if no column of the board is covered by two checkers, there are $\binom{n}{4}$ ways to choose the four columns containing one checker and then $\binom{4}{2} = 6$ ways to choose which columns have a checker in the upper row. This gives $\binom{n}{4}\binom{4}{2} = 6\binom{n}{4}$ arrangements with no entirely covered column.

2.7.13. Give a combinatorial proof that $\sum_{r=1}^{n} r = \left(\!\binom{n}{2}\!\right)$.

Answer

Suppose there are n flavors at the ice cream store, numbered 1 through n. There are then $\left(\!\binom{n}{2}\!\right)$ ways to buy 2 cones, where a flavor can be repeated. Alternatively, suppose that the larger flavor is r, so there are r ways to choose the other cone from flavors 1 through r. Summing over r from 1 through n also shows there are $\sum_{r=1}^{n} r$ ways to buy two cones.

3

BINOMIAL SERIES AND GENERATING FUNCTIONS

PROBLEM SET 3.2

3.2.1. Evaluate each of these sums:

(a) $\sum_{n=0}^{8} \binom{8}{n} 9^n$ **(b)** $\sum_{n=0}^{8} (-1)^n \binom{8}{n} 11^n$

Answer

(a) $\sum_{n=0}^{8} \binom{8}{n} 9^n = (1+9)^8 = 10^8$ **(b)** $\sum_{n=0}^{8} (-1)^n \binom{8}{n} 11^n = (1-11)^8 = (-10)^8 = 10^8$

3.2.3. Determine the coefficient of w^3x^4 in the expansion of each of these binomials:

(a) $(w+2x)^7$ **(b)** $\left(2w-x^2\right)^7$ **(c)** $\left(2w-x^2\right)^5$

Answer

(a) $\binom{7}{3} 2^4 = 35 \cdot 16 = 560$

(b) 0, since no term involving w^3x^4 appears in the expansion

(c) $\binom{5}{3} \left(2^3\right)(-1)^2 = 80$

Solutions Manual to Accompany Combinatorial Reasoning: An Introduction to the Art of Counting, First Edition. Duane DeTemple and William Webb.

3.2.5. Prove algebraically that

(a) $\binom{n}{r} = \binom{n-2}{r-2} + 2\binom{n-2}{r-1} + \binom{n-2}{r}$

(b) $\binom{n}{r} = \binom{n-3}{r-3} + 3\binom{n-3}{r-2} + 3\binom{n-3}{r-1} + \binom{n-3}{r}$

Answer

(a) $\binom{n}{r} = \binom{n-1}{r-1} + \binom{n-1}{r} = \binom{n-2}{r-2} + \binom{n-2}{r-1} +$ $\binom{n-2}{r-1} + \binom{n-2}{r}$ using Pascal's identity twice.

(b) By part (a) and Pascal's identity,

$$\binom{n}{r} = \binom{n-2}{r-2} + 2\binom{n-2}{r-1} + \binom{n-2}{r} = \binom{n-3}{r-3} + \binom{n-3}{r-2}$$
$$+2\binom{n-3}{r-2} + 2\binom{n-3}{r-1} + \binom{n-3}{r-1} + \binom{n-3}{r}$$

which simplifies to the desired expression.

3.2.7. Give a combinatorial proof of the identity

$$0 = \sum_{k=0}^{n} (-1)^k \binom{n}{k}, n \geq 0$$

[*Hint*: Pair subsets of $[n]$ that differ only by the inclusion or exclusion of the element 1.]

Answer

Every subset S of $[n]$ can be uniquely paired to the subset S' obtained by either deleting element 1 from S if $1 \in S$ or by inserting element 1 into S if $1 \notin S$. Since $|S|$ and $|S'|$ have opposite parity—one is even and the other is odd—the number of subsets of $[n]$ with an even number of elements is the same as the number of subsets with an odd number of elements. That is, $\sum_{r\,\text{odd}} \binom{n}{r} = \sum_{r\,\text{even}} \binom{n}{r}$, which is equivalent to the identity.

3.2.9. **(a)** Use the binomial theorem to prove that

$$\sum_{r=1}^{n} r\binom{n}{r} = n2^{n-1}$$

(b) Show from part (a) that the average number of elements in the subsets of an n-element set is $n/2$.

(c) Give a combinatorial reason for the result of part (b). [*Hint*: What is a way to pair up the subsets of the n-set?]

Answer

(a) Differentiating $(1+x)^n = \sum_{r=0}^{n} \binom{n}{r} x^r$ gives $n(1+x)^{n-1} = \sum_{r=1}^{n} r \binom{n}{r} x^{r-1}$. Setting $x = 1$ gives the result.

(b) There are 2^n subsets of an n element set, and $\binom{n}{r}$ of these subsets have r elements. Thus the average number of elements of all of the subsets is $\dfrac{\sum_{r=0}^{n} r \binom{n}{r}}{2^n} = \dfrac{n2^{n-1}}{2^n} = \dfrac{n}{2}$.

(c) Pair each subset S with its complement $\overline{S}$. If $|S| = r$ then $|\overline{S}| = n - r$. The average number of elements in each complementary pair is therefore $n/2$, so this is also the average over all of the complementary pairs.

3.2.11. Prove that for all $n \geq 3$, $0 = \sum_{r=1}^{n-1} (-1)^r (n-r) r \binom{n}{r}$.

Answer

Choose $w = 1$ and $x = -1$ in the identity proved in part (a) of Problem 3.2.10.

3.2.13. Expand these multinomials.

(a) $(r - 2s + t)^3$ **(b)** $(w + x + 2y - 3z)^2$

Answer

(a) $(r-2s+t)^3 = \binom{3}{3,0,0} r^3 + \binom{3}{2,1,0} r^2(-2s) + \binom{3}{2,0,1} r^2 t + \binom{3}{1,2,0} r(-2s)^2 + \binom{3}{1,1,1} r(-2s)t + \binom{3}{1,0,2} rt^2 + \binom{3}{0,3,0} (-2s)^3 + \binom{3}{0,2,1} (-2s)^2 t + \binom{3}{0,1,2} (-2s) t^2 + \binom{3}{0,0,3} t^3 = r^3 - 6r^2 s + 3r^2 t + 12rs^2 - 12rst + 3rt^2 - 8s^3 + 12s^2 t - 6st^2 + t^3$

(b) $(w + x + 2y - 3z)^2 = w^2 + x^2 + 4y^2 + 9z^2 + 2wx + 4wy - 6wz + 4xy - 6xz - 12yz$

3.2.15. What is the coefficient of x^2y^4z in:
(a) $(x+y+z)^7$ **(b)** $(2x+3y-z)^7$ **(c)** $(x+2y+z+1)^7$

Answer

(a) $\binom{7}{2,4,1} = \frac{7!}{2!4!1!} = 105$

(b) $\binom{7}{2,4,1} 2^2 3^4 (-1)^1 = -34{,}020$

(c) $\binom{7}{2,4,1,0} 2^4 = 1680$

3.2.17. Prove that

$$\binom{n}{k} k^{n-k} = \frac{1}{k!} \sum_{\substack{r_1,r_2,\ldots,r_k \geq 0 \\ r_1+r_2+\cdots+r_k=n}} r_1 r_2 \cdots r_k \binom{n}{r_1, r_2, \ldots, r_k}$$

Answer

Take the k-fold derivative $\frac{\partial^k}{\partial x_1 \partial x_2 \cdots \partial x_k}$ of the multinomial formula and then set $x_1 = x_2 = \cdots = x_k = 1$.

PROBLEM SET 3.3

3.3.1. Determine the generating functions of these sequences.
(a) $a_n = 2^n, n = 0, 1, 2, \ldots$
(b) $b_n = 3(-4)^n, n = 0, 1, 2, \ldots$
(c) $c_n = 2^n - 3(-4)^n, n = 0, 1, 2, \ldots$

Answer

(a) $\sum_{n\geq 0} 2^n x^n = \sum_{n\geq 0} (2x)^n = \frac{1}{1-2x}$

(b) $\sum_{n\geq 0} 3(-4)^n x^n = 3\sum_{n\geq 0} (-4x)^n = \frac{3}{1+4x}$

(c) Using parts (a) and (b), the OGF is
$\frac{1}{1-2x} - \frac{3}{1+4x} - = \frac{10x-2}{1+2x-8x^2}$

3.3.3. What sequences are generated by these OGFs?
(a) $f(x) = \frac{1}{1-3x}$ **(b)** $g(x) = \frac{1}{2x-2}$ **(c)** $h(x) = \frac{1+x}{2-8x+6x^2}$

[*Hint*: Parts (a) and (b) will be helpful for part (c) once $h(x)$ is written in partial fraction form.]

Answer

(a) $a_n = 3^n$

(b) $b_n = -1/2$

(c) $h(x) = \dfrac{1+x}{2-8x+6x^2} = \dfrac{1}{1-3x} - \dfrac{1}{2-2x} = f(x) + g(x)$ so the sequence is $c_n = a_n + b_n = 3^n - \dfrac{1}{2}$.

3.3.5. **(a)** Use differentiation to show that

$$\frac{x}{(1-x)^2} = \sum_{k=0}^{\infty} kx^k$$

assuming you already know the geometric series expansion given by (3.2).

(b) What sequence has the OGF $\dfrac{x^3}{(1-5x)^2}$?

(c) What is the OGF of the sequence $0, 6, 24, 72, \ldots, 3n2^n, \ldots$?

Answer

(a) The derivative of the geometric series is $\dfrac{1}{(1-x)^2} = \sum_{k=1}^{\infty} kx^{k-1}$, so the desired formula is obtained by a multiplication by x.

(b) $\dfrac{x^3}{(1-5x)^2} = \dfrac{x^2}{5}\dfrac{(5x)}{(1-(5x))^2} = \dfrac{x^2}{5}\sum_{k=1}^{\infty} k(5x)^k = \sum_{k=1}^{\infty} k5^{k-1}x^{k+2} =$ $x^3 + 10x^4 + 75x^5 + \cdots$ so the sequence is $a_0 = a_1 = a_2 = 0$, $a_k = (k-2)\,5^{k-3}, k \geq 3$.

(c) $\sum_{k=0}^{\infty} 3k2^k x^k = 3\sum_{k=0}^{\infty} k(2x)^k = \dfrac{3(2x)}{(1-(2x))^2} = \dfrac{6x}{(1-2x)^2}$

3.3.7. The *Fibonacci Sequence OGF* is obtained as follows. Suppose that $x^2 = 1 + x$ has the roots φ and $\hat{\varphi}$, so that $x^2 - x - 1 = (x-\varphi)(x-\hat{\varphi})$.

(a) Show that $\varphi + \hat{\varphi} = 1$ and $\varphi\hat{\varphi} = -1$.

[*Hint*: Equate coefficients of x.]

(b) Show that

$$\frac{x}{1-x-x^2} = \frac{1}{\varphi - \hat{\varphi}}\left(\frac{1}{1-\varphi x} - \frac{1}{1-\hat{\varphi}x}\right).$$

(c) Show that

$$\frac{x}{1-x-x^2} = \sum_{n=0}^{\infty} \widehat{F}_n x^n \text{ where } \widehat{F}_n = \frac{\varphi^n - \hat{\varphi}^n}{\varphi - \hat{\varphi}}.$$

(d) Verify that $\widehat{F_n} = F_n$ is the nth Fibonacci number, by showing that

$$\widehat{F}_0 = 0, \widehat{F}_1 = 1 \text{ and } \widehat{F}_n = \widehat{F}_{n-1} + \widehat{F}_{n-2}.$$

[*Hint*: $\varphi^2 = \varphi + 1$ and $\hat{\varphi}^2 = \hat{\varphi} + 1$.]

(e) Show that

$$\widehat{F}_n = F_n = \frac{1}{\sqrt{5}}\left(\left(\frac{1+\sqrt{5}}{2}\right)^n - \left(\frac{1-\sqrt{5}}{2}\right)^n\right)$$

where F_n is the nth Fibonacci number.

[*Note*: We have shown that the OGF of the Fibonacci sequence is $\dfrac{x}{1-x-x^2} = \sum_{n=0}^{\infty} F_n x^n$. The formula for the Fibonacci numbers given in part (e) is known as *Binet's formula*.]

(f) Show that $F_n = \left\{\dfrac{\varphi^n}{\sqrt{5}}\right\}$, $n \geq 1$, where $\{x\}$ is the nearest integer to x function.

[*Hint*: $|\hat{\varphi}| = \dfrac{\sqrt{5}-1}{2} = 0.618\ldots < 1$ and $\sqrt{5} > 2$.]

Answer

(a) Since $(x-\varphi)(x-\hat{\varphi}) = x^2 - (\varphi + \hat{\varphi})x + \varphi\hat{\varphi} = x^2 - x - 1$, equating coefficients gives the result.

(b) Using the results of part (a), $\dfrac{1}{\varphi - \hat{\varphi}}\left(\dfrac{1}{1-\varphi x} - \dfrac{1}{1-\hat{\varphi} x}\right) = \dfrac{1}{\varphi - \hat{\varphi}}\left(\dfrac{1-\hat{\varphi}x - 1 + \varphi x}{(1-\varphi x)(1-\hat{\varphi} x)}\right) = \dfrac{1}{\varphi - \hat{\varphi}}\left(\dfrac{(\varphi - \hat{\varphi})x}{1-(\varphi + \hat{\varphi})x - x^2}\right) = \dfrac{x}{1-x-x^2}$.

(c) $\dfrac{1}{\varphi - \hat{\varphi}}\left(\dfrac{1}{1-\varphi x} - \dfrac{1}{1-\hat{\varphi} x}\right) = \dfrac{1}{\varphi - \hat{\varphi}} \sum_{n=0}^{\infty} (\varphi^n - \hat{\varphi}^n) x^n.$

(d) $\widehat{F}_0 = \dfrac{\varphi^0 - \hat{\varphi}^0}{\varphi - \hat{\varphi}} = \dfrac{0}{\varphi - \hat{\varphi}} = 0, \widehat{F}_1 = \dfrac{\varphi^1 - \hat{\varphi}^1}{\varphi - \hat{\varphi}} = 1$ and $\widehat{F}_{n-1} + \widehat{F}_{n-2} =$
$\dfrac{\varphi^{n-1} - \hat{\varphi}^{n-1} + \varphi^{n-2} - \hat{\varphi}^{n-2}}{\varphi - \hat{\varphi}} = \dfrac{\varphi^{n-2}(\varphi + 1) - \hat{\varphi}^{n-2}(\hat{\varphi} + 1)}{\varphi - \hat{\varphi}} =$
$\dfrac{\varphi^{n-2}\varphi^2 - \hat{\varphi}^{n-2}\hat{\varphi}^2}{\varphi - \hat{\varphi}} = \dfrac{\varphi^n - \hat{\varphi}^n}{\varphi - \hat{\varphi}} = \widehat{F}_n.$

(e) Part (d) shows that $\widehat{F}_n = F_n$. Using the quadratic formula, the roots of $x^2 - x - 1 = 0$ are $\varphi = \dfrac{1 + \sqrt{5}}{2}$ and $\hat{\varphi} = \dfrac{1 - \sqrt{5}}{2}$, so Binet's formula follows from part (c).

(f) $\left| F_n - \dfrac{\varphi^n}{\sqrt{5}} \right| = \left| \dfrac{\hat{\varphi}^n}{\sqrt{5}} \right| \le \left| \dfrac{\hat{\varphi}}{\sqrt{5}} \right| < \dfrac{0.7}{2} < \dfrac{1}{2}, n \ge 1$, so the integer nearest $\dfrac{\varphi^n}{\sqrt{5}}$ is F_n for $n \ge 1$.

3.3.9. Use differentiation to give an induction proof of $(1 - x)^{-n} = \sum_{k \ge 0} \binom{k + n - 1}{n - 1} x^k$, assuming you already know the geometric series $(1 - x)^{-1} = \sum_{k \ge 0} x^k$, which is the base case of the induction.

Answer

The derivative of $(1 - x)^{-n} = \sum_{k \ge 0} \binom{k + n - 1}{n - 1} x^k$ is $n(1 - x)^{-n-1} =$
$\sum_{k \ge 1} k \binom{k + n - 1}{n - 1} x^{k-1} = \sum_{k \ge 1} \dfrac{k(k + n - 1)!}{k!(n - 1)!} x^{k-1} = \sum_{k \ge 1} \dfrac{n(k + n - 1)!}{(k - 1)!n!} x^{k-1}$
$= n \sum_{k \ge 0} \binom{k + n}{n} x^k$ where $k - 1$ is replaced by k in the last equality.

Thus we see that $(1 - x)^{-(n+1)} = \sum_{k \ge 0} \binom{k + n}{n} x^k$, so the formula holds in the case $n + 1$ when it holds in case n. This completes the proof by mathematical induction.

3.3.11. Prove that
(a) $(-1)_n = (-1)^n n!$ **(b)** $(-\alpha)_n = (-1)^n (\alpha - n + 1)_n$

Answer

(a) $(-1)_n = (-1)(-2) \cdots (-1 - n + 1) = (-1)^n (1)(2) \cdots (n) = (-1)^n n!$

(b) $(-\alpha)_n = (-\alpha)(-\alpha - 1) \cdots (-\alpha - n + 1) =$
$(-1)^n (\alpha)(\alpha + 1) \cdots (\alpha + n - 1) = (-1)^n (\alpha + n - 1)_n$

3.3.13. Evaluate $\binom{\frac{1}{3}}{n}$.

Answer

$$\frac{\left(\frac{1}{3}\right)\left(-\frac{2}{3}\right)\left(-\frac{5}{3}\right)\left(-\frac{8}{3}\right)\cdots\left(\frac{1}{3}-n+1\right)}{n!} = \frac{(-1)^{n-1}\,2\cdot 5\cdot 8\cdot \cdots \cdot (3n-4)}{3^n n!}$$

3.3.15. Use four terms of the binomial series to derive the approximation $\frac{1}{\sqrt{1.1}} \doteq 0.9534375$. How many digits beyond the decimal point are correct?

Answer

$$(1+0.1)^{-1/2} \doteq \sum_{k=0}^{3} \frac{(-1)^k}{2^{2k}}\binom{2k}{k}\frac{1}{10^k} = 1 - \frac{1}{2^2}\binom{2}{1}\frac{1}{10} + \frac{1}{2^4}\binom{4}{2}\frac{1}{10^2}$$
$$-\frac{1}{2^6}\binom{6}{3}\frac{1}{10^3} = 1 - \frac{1}{20} + \frac{6}{1600} - \frac{20}{64000} = 0.9534375$$

Since $\frac{1}{\sqrt{1.1}} = 0.95346\ldots$, the first 4 digits following the decimal are correct.

3.3.17. Prove that $\frac{1}{\sqrt{1-4x}} = \sum_{k\geq 0}\binom{2k}{k}x^k$.

Answer
Replacing x with $-4x$ in expansion (3.34) gives us

$$\frac{1}{\sqrt{1-4x}} = \sum_{k\geq 0}\frac{(-1)^k}{2^{2k}}\binom{2k}{k}(-4)^k x^k = \sum_{k\geq 0}\binom{2k}{k}x^k.$$

PROBLEM SET 3.4

3.4.1. **(a)** Show that the OGF of the sequence of squares $1, 2^2, 3^2, \ldots, n^2, \ldots$ is $\sum_{k=1}^{\infty} k^2 x^k = (xD)^2\,\frac{1}{1-x} = \frac{x+x^2}{(1-x)^3}$ where $D = \frac{d}{dx}$.

(b) Show that the OGF for the sequence of cubes $1, 2^3, 3^3, \ldots, k^3, \ldots$ is $\sum_{k=0}^{\infty} k^3 x^k = (xD)^3\,\frac{1}{1-x} = \frac{x\left(x^2+4x+1\right)}{(1-x)^4}$.

Answer

(a)

$$\sum_{k=1}^{\infty} k^2 x^k = (xD)^2 \frac{1}{1-x} = (xD)\, x\,(1-x)^{-2}$$

$$= x\left[(1-x)^{-2} + 2x\,(1-x)^{-3}\right] = \frac{x\,[1-x+2x]}{(1-x)^3} = \frac{x+x^2}{(1-x)^3}$$

(b)

$$\sum_{k\geq 0} k^3 x^k = (xD)^3 \frac{1}{1-x} = (xD)\left(x+x^2\right)(1-x)^{-3}$$

$$= x\left[(1+2x)(1-x)^{-3} + 3\left(x+x^2\right)(1-x)^{-4}\right]$$

$$= \frac{x}{(1-x)^4}\left[(1+2x)(1-x) + 3\left(x+x^2\right)\right]$$

$$= \frac{x}{(1-x)^4}\left[1+x-2x^2+3x+3x^2\right] = \frac{x\left(x^2+4x+1\right)}{(1-x)^4}$$

3.4.3. **(a)** Verify that $k^3 - (k-1)^3 = 3k^2 - 3k + 1$.

(b) Sum the result of part (a) over $k = 1, 2, \ldots, n$ and obtain a new derivation of the identity $1^2 + 2^2 + \cdots + n^2 = \dfrac{n\,(n+1)\,(2n+1)}{6}$.

Answer

(a) $(k-1)^3 = k^3 - 3k^2 + 3k - 1$ by the binomial theorem.

(b) $n^3 = \sum_{k=1}^{n}\left(k^3 - (k-1)^3\right) = 3\sum_{k=1}^{n} k^2 - 3\sum_{k=1}^{n} k + \sum_{k=1}^{n} 1 = 3\sum_{k=1}^{n} k^2 - 3\dfrac{n\,(n+1)}{2} + n$

$$\therefore \sum_{k=1}^{n} k^2 = \frac{n^3}{3} + \frac{n^2+n}{2} - \frac{n}{3} = \frac{2n^3+3n^2+n}{6} = \frac{n\,(n+1)\,(2n+1)}{6}$$

3.4.5. A *centered polygonal number* is the number of dots in a pattern of polygons that surround a central dot and continue to increase the number of dots along each edge of the subsequent polygons. For example, the centered 7-gonal numbers are $c_0^{(7)} = 1, c_1^{(7)} = 8, c_2^{(7)} = 22, c_3^{(7)} = 43$.

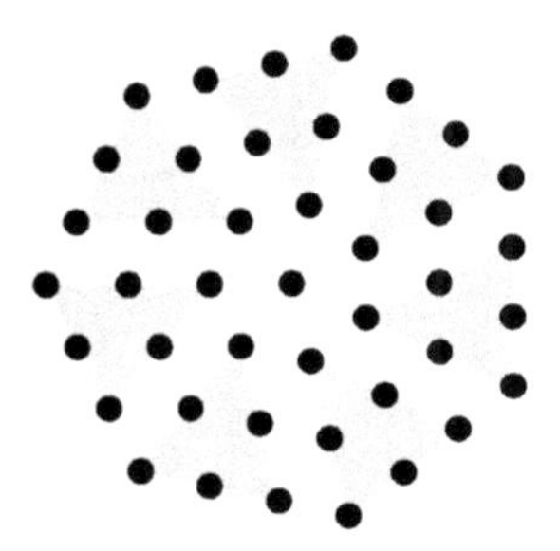

The shading shows that $c_3^{(7)} = 1 + 7t_3 = 1 + 7\left(\frac{4 \cdot 3}{2}\right) = 43$, where $t_j = \frac{1}{2}(j+1)j$ is the jth triangular number. More generally, the kth centered r-gonal number $c_k^{(r)}$ is given by $c_k^{(r)} = 1 + rt_k = 1 + r\binom{k+1}{2}$, $k \geq 0$.

(a) Show that the generating function for the sequence of centered r-gonal numbers is

$$g^{(r)}(x) = \frac{1 + (r-2)x + x^2}{(1-x)^3}.$$

(b) Use the OGF obtained in (a) to verify that $c_4^{(10)} = 101$.

Answer

(a) Since

$$c_0^{(r)} = 1, g^{(r)}(x) = \sum_{k \geq 0} c_k^{(r)} x^k = \sum_{k \geq 0} \left(1 + r\binom{k+1}{2}\right) x^k$$

$$= \sum_{k \geq 0} x^k + r \sum_{k \geq 0} \binom{k+1}{2} x^k = \frac{1}{1-x} + r\frac{x}{(1-x)^3} = \frac{1 + (r-2)x + x^2}{(1-x)^3}$$

(b) $[x^4]\,\frac{1 + 8x + x^2}{(1-x)^3} = [x^4]\left(1 + 8x + x^2\right) \sum_{k \geq 0} \binom{k+2}{2} x^k = \binom{6}{2} + 8\binom{5}{2} + \binom{4}{2} = 15 + 80 + 6 = 101.$

3.4.7. Find and simplify the OGF for the number of solutions in integers of

$$x_1 + x_2 + x_3 = n, x_1 \geq 0, x_2 \geq 3, 4 \geq x_1 \geq 1.$$

Answer

$$\left(1+x+x^2+\cdots\right)\left(x^3+x^4+x^5+\cdots\right)\left(x+x^2+x^3+x^4\right)=\frac{x^4\left(1-x^4\right)}{(1-x)^3}$$

3.4.9. **(a)** Determine the form of an OGF for the number of solutions of the equation $a_1 + a_2 + a_3 = n$, where a_1 is a positive square, a_2 is a positive cube, and a_3 is a prime number. Don't attempt to simplify your expression.

(b) Find the number of solutions when $n = 12$.

(c) Use a computer (CAS) to find the number of solutions when $n = 46$.

Answer

(a) $\left(x+x^4+x^9+\cdots\right)\left(x+x^8+x^{27}+\cdots\right)\left(x^2+x^3+x^5+x^7+x^{11}+\cdots\right)$

(b) Three solutions

(c) Seven solutions

3.4.11. Obtain, as a product of series, the generating functions for the number of solutions in integers of the equation $2x_1 + 4x_2 + 5x_3 = n, \quad n \geq 0$, where one of these conditions hold:

(a) x_1, x_2, x_3 are nonnegative integers (b) $x_1 \geq 3, x_2 \geq 2, x_3 \geq 1$

Answer

(a) $f(x) = \left(1+x^2+x^4+\cdots\right)\left(1+x^4+x^8+\cdots\right)\left(1+x^5+x^{10}+\cdots\right) = \dfrac{1}{\left(1-x^2\right)\left(1-x^4\right)\left(1-x^5\right)}$

(b) $f(x) = \left(x^6+x^8+\cdots\right)\left(x^8+x^{12}+\cdots\right)\left(x^5+x^{10}+\cdots\right) = \dfrac{x^{19}}{\left(1-x^2\right)\left(1-x^4\right)\left(1-x^5\right)}$

3.4.13. The Chinese game mahjong is often played with 144 tiles. There are 8 distinct tiles called *flowers* and *seasons*, and 4 copies each of 34 other kinds of tiles. A *hand* consists of 13 tiles.

(a) Write a generating function for the number of Mahjong hands and then determine the number of hands using a CAS.

(b) How many hands do not contain either a flower or season? Give an answer in terms of binomial coefficients, and determine the numerical value with a CAS.

(c) Points are awarded for having either a pung—three of a kind—or a kong—four of a kind. What is the probability that a hand has at least one kong?

(d) How many hands contain neither a pung nor a kong?

Answer

(a) The number of hands is $\left[x^{13}\right](1+x)^8\left(1+x+x^2+x^3+x^4\right)^{34} =$ 678,524,880,012

(b)

$$\begin{aligned}\left[x^{13}\right]\left(1+x+x^2+x^3+x^4\right)^{34} &= \left[x^{13}\right]\left(\frac{1-x^5}{1-x}\right)^{34} \\ &= \left[x^{13}\right](1-34x^5+\binom{34}{2}x^{10}+\cdots)\sum_{k=0}^{\infty}\left(\!\binom{34}{k}\!\right)x^k \\ &= \left(\!\binom{34}{13}\!\right) - 34\left(\!\binom{34}{8}\!\right) + \frac{34\cdot 33}{2}\left(\!\binom{34}{3}\!\right) \\ &= \binom{46}{13} - 34\binom{41}{8} + 561\binom{36}{3} \\ &= 101766230790 - 3248640330 + 4005540 \\ &= 98{,}521{,}596{,}000\end{aligned}$$

(c) The number of hands without a kong is $\left[x^{13}\right](1+x)^8(1+x+x^2+x^3)^{34} = 625{,}342{,}882{,}242$, so there are $678524880012 - 625342882242 = 53181997770$ hands with at least one kong. The probability of this event is $53181997770/678524880012 \doteq 0.078$.

(d) $\left[x^{13}\right](1+x)^8\left(1+x+x^2\right)^{34} = 403{,}867{,}704{,}736$

3.4.15. Pólya asked for the number of ways to make change of an American dollar using coins—pennies, nickels, dimes, quarters, and half-dollars.

(a) Give the OGF $h(x)$ for which the answer to Pólya's question is given by the expression $\left[x^{100}\right]h(x)$.

(b) If you have access to Maple, Mathematica, MatLab, or another CAS, show that there are 292 ways to make change for a dollar bill in coins of smaller denomination. For example, in Maple, use the command

```
coeff(series(((1-x^5)*(1-x^10)*(1-x^25)*(1-x^50)
*(1-x))^(-1),(x=0,101)),x^100);
```

Answer

(a)

$$\begin{aligned}h(x) &= f_P(x)f_N(x)f_D(x)f_Q(x)f_H(x) \\ &= \left(\sum_{r\geq 0}x^r\right)\left(\sum_{s\geq 0}x^{5s}\right)\left(\sum_{t\geq 0}x^{10t}\right)\left(\sum_{u\geq 0}x^{25u}\right)\left(\sum_{v\geq 0}x^{50v}\right) \\ &= \frac{1}{1-x}\frac{1}{1-x^5}\frac{1}{1-x^{10}}\frac{1}{1-x^{25}}\frac{1}{1-x^{50}}\end{aligned}$$

(b) The command > coeff(series(((1-x^5)*(1-x^10)*(1-x^25)*(1-x^50)*(1-x))^(-1),(x=0,101)),x^100);
returns 292.

3.4.17. Give a new derivation of the closed form of the generating function

$$f_n(x) = \sum_{k \geq 0} \left(\!\binom{n}{k}\!\right) x^k$$

by using this model: let $f_n(x)$ be the generating function for the number of ways to buy k cones at an ice cream parlor that offers n flavors. Now justify these steps.

(a) If only vanilla is available, then $f_1(x) = \dfrac{1}{1-x}$.

(b) If vanilla and n other flavors are available, then $f_{n+1}(x) = f_n(x) f_1(x)$.

(c) $f_n(x) = \sum_{k \geq 0} \left(\!\binom{n}{k}\!\right) x^k = \dfrac{1}{(1-x)^n}$

Answer

(a) There is just one way to buy k vanilla cones, so $f_1(x) = \sum_{k \geq 0} 1 \cdot x^k = \dfrac{1}{1-x}$.

(b) Any order of cones can be viewed as a process with two types of steps: A—how many cones are vanilla, and B—how many cones are not vanilla. Thus, $f_{n+1}(x) = f_n(x) f_1(x)$.

(c) By iteration, $f_n(x) = f_{n-1}(x) f_1(x) = f_{n-2}(x) \left(f_1(x)\right)^2 = \cdots = \left(f_1(x)\right)^n = \dfrac{1}{(1-x)^n}$.

3.4.19. Starting with the OGF

$$g(x) = \frac{1}{(1-x)^{r+1}} = \sum_{k \geq 0} \binom{k+r}{r} x^k$$

prove the hockey stick identity

$$\binom{n+1}{r+1} = \binom{r}{r} + \binom{r+1}{r} + \cdots + \binom{n}{r}.$$

Answer
$t_{n-r} = \sum_{k=0}^{n-r} \binom{k+r}{r}$ is the $(n-\text{r})^{\text{th}}$ partial sum sequence of the coefficients of the OGF $g(x)$, so

$$t_{n-r} = \left[x^{n-r}\right] \frac{1}{1-x} g(x) = \left[x^n\right] \frac{x^r}{(1-x)^{r+2}}$$
$$= \left[x^n\right] \sum_{k\geq 0} \binom{k+r+1}{r+1} x^{k+r} = \binom{n+1}{r+1}.$$

3.4.21. In Example 3.32, suppose that Kerry has six eggs to color, an odd number of which are to be dyed blue, either one or two to be dyed red, and the rest dyed green. What are his coloring choices?

Answer
The five ways are 5B+1R, 3B+2R+1G, 3B+1R+2G, 1B+2R+3G, 1B+1R+4G.

3.4.23. A $1 \times n$ board will be tiled with red or green unit squares at the left, then by 1×2 dominoes, and finished with unit squares that include exactly one black tile and the rest white tiles. One possible tiling of a board of length 10 is the RRGDDWBW tiling shown here:

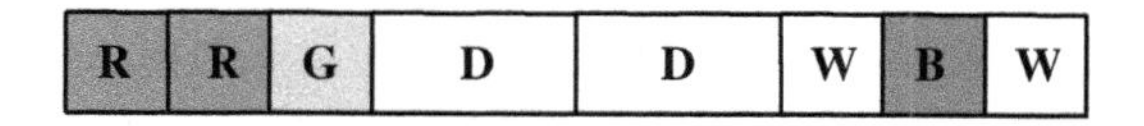

(a) In how many ways, h_n, can the board of length n be tiled?

(b) Verify your formula in part (a) by listing the 12 tilings of a board of length 3.

Answer

(a) If r squares are to be used at the left end of the board, there are 2^r ways to choose the colors of the squares. This gives the OGF $f_A(x) = \sum_{r\geq 0} 2^r x^r = \sum_{r\geq 0} (2x)^r = \frac{1}{1-2x}$. Since dominoes are of length 2, s dominoes cover $2s$ cells of the board, and the OGF is therefore $f_B(x) = \sum_{s\geq 0} x^{2s} = \frac{1}{1-x^2}$. Finally, if $t \geq 1$ unit squares are used at the right of the board, there are t ways to choose the one that is black. This gives the OGF $f_C(x) = \sum_{t\geq 0} tx^t = x \sum_{t\geq 1} tx^{t-1} =$

$x\frac{d}{dx}\sum_{t\geq 0} x^t = x\frac{d}{dx}\frac{1}{1-x} = \frac{x}{(1-x)^2}$. The OGF for the number of tilings of boards of any length is therefore

$$\begin{aligned}
h(x) &= f_A(x) f_B(x) f_C(x) \\
&= \frac{1}{1-2x}\frac{1}{1-x^2}\frac{x}{(1-x)^2} \\
&= \frac{x}{(1-2x)(1+x)(1-x)^3} \\
&= \frac{8}{3(1-2x)} - \frac{1}{24(1+x)} - \frac{11}{8(1-x)} - \frac{3}{4(1-x)^2} - \frac{1}{2(1-x)^3},
\end{aligned}$$

where a CAS has given the partial fraction decomposition. Thus the number of ways to tile a board of length n is

$$\begin{aligned}
h_n &= [x^n]\, h(x) \\
&= [x^n]\left[\frac{8}{3}\sum_{n\geq 0} 2^n x^n - \frac{1}{24}\sum_{n\geq 0}(-1)^n x^n - \frac{11}{8}\sum_{n\geq 0} x^n \right. \\
&\quad \left. - \frac{3}{4}\sum_{n\geq 0}(n+1)x^n - \frac{1}{4}\sum_{n\geq 0}(n+1)(n+2)x^n\right] \\
&= \frac{2^{n+3}}{3} - \frac{(-1)^n}{24} - \frac{11}{8} - \frac{3(n+1)}{4} - \frac{(n+1)(n+2)}{4} \\
&= \frac{2^{n+6} - (-1)^n - 33 - 6(n+1)(n+5)}{24}.
\end{aligned}$$

(b) DB, RRB, RGB, GRB, GGB, WWB, WBW, BWW, GBW, GWB, RBW, RWB.

PROBLEM SET 3.5

3.5.1. Let $h(x) = \sum_{n\geq 0} h_n x^n$ be a power series. Verify that $\left[\frac{x^n}{n!}\right] f(x) = n!\,[x^n] f(x)$.

Answer

$$\left[\frac{x^n}{n!}\right] f(x) = \left[\frac{x^n}{n!}\right]\sum_{m\geq 0} h_m x^m = \left[\frac{x^n}{n!}\right]\sum_{m\geq 0} m!h_m\left(\frac{x^m}{m!}\right) = n!h_n = n!\,[x^n] f(x)$$

3.5.3. Give the EGF that corresponds to each of these sequences.

(a) $1, 0, -1, 0, 1, 0, -1, 0, \ldots$

(b) $0, 1, 0, -1, 0, 1, 0, -1, 0, \ldots$

[*Hint*: Both EGFs are a common trigonometric function.]

Answer

(a) $\sum_{n=0}^{\infty}(-1)^n\frac{x^{2n}}{(2n)!} = \cos x$

(b) $\sum_{n=0}^{\infty}(-1)^n\frac{x^{2n+1}}{(2n+1)!} = \sin x$

3.5.5. Let $f^{(e)}(x) = \sum_{n\geq 0} a_n\frac{x^n}{n!}$ be the EGF of the sequence $a_0, a_1, a_2, \ldots, a_n, \ldots$.

Verify that

(a) $f^{(e)\prime}(x) = \sum_{n\geq 0} a_{n+1}\frac{x^n}{n!}$ is the EGF of the sequence $a_1, a_2, \ldots, a_{n+1}, \ldots$.

(b) $xf^{(e)}(x)$ is the EGF of the sequence $0, a_0, 2a_1, 3a_2, \ldots, na_{n-1}, \ldots$.

(c) $\left[\frac{x^n}{n!}\right]x^r f^{(e)}(x) = \begin{cases} 0, & 0 \leq n < r \\ (n)_r\, a_{n-r}, & r \geq n \end{cases}.$

Answer

(a) $f^{(e)\prime}(x) = \frac{d}{dx}\sum_{n\geq 0} a_n\frac{x^n}{n!} = \sum_{n\geq 1} a_n\frac{nx^{n-1}}{n!} = \sum_{n\geq 1} a_n\frac{x^{n-1}}{(n-1)!} =$

$\sum_{n\geq 0} a_{n+1}\frac{x^n}{n!}$

(b) $xf^{(e)}(x) = \sum_{n=0}^{\infty} a_n\frac{x^{n+1}}{n!} = \sum_{n=0}^{\infty}(n+1)\,a_n\frac{x^{n+1}}{(n+1)!} = 0 + \sum_{n=1}^{\infty} na_{n-1}\frac{x^n}{n!}$

(c) $x^r f^{(e)}(x) = \sum_{n=0}^{\infty} a_n\frac{x^{n+r}}{n!} = \sum_{n=r}^{\infty} a_{n-r}\frac{x^n}{(n-r)!} = \sum_{n=r}^{\infty} a_{n-r}\frac{n!}{(n-r)!}\frac{x^n}{n!}$

$$= 0 + 0\frac{x}{1!} + \cdots + 0\frac{x^{r-1}}{(r-1)!} + \sum_{n=r}^{\infty} a_{n-r}\,(n)_r\,\frac{x^n}{n!}$$

3.5.7. Let h_n denote the number of ways that a $1 \times n$ rectangle can be tiled with red, blue, green, and yellow squares, where there must be an even number of red tiles, an even number of blue tiles, and an least two green tiles.

(a) Find a formula for h_n.

(b) Use the formula to determine h_3 and then list the tilings of the 1×3 board.

Answer

(a)

$$\left[\frac{x^n}{n!}\right]\left(\frac{e^x+e^{-x}}{2}\right)^2\left(e^x-1-x\right)e^x=\left[\frac{x^n}{n!}\right]\frac{\left(e^{2x}+2+e^{-2x}\right)\left(e^{2x}-e^x-xe^x\right)}{4}$$

$$=\left[\frac{x^n}{n!}\right]\frac{\left(e^{4x}-e^{3x}-xe^{3x}+2e^{2x}-2e^x-2xe^x+1-e^{-x}-xe^{-x}\right)}{4}$$

$$=\frac{4^n-3^n-n3^{n-1}+2^{n+1}-2-2n-(-1)^n-(-1)^{n-1}n}{4}$$

(b) $h_3=\frac{1}{4}(64-27-27+16-2-6+1-3)=4$: GGG, GGY, GYG, YGG

3.5.9. Find the EGF that can be used to find the number of ways to distribute 10 distinct candy bars to 4 children, where the oldest child must receive either, 2, 3, 5 or 8 candy bars.

Answer

$$\left(\frac{x^2}{2!}+\frac{x^3}{3!}+\frac{x^5}{5!}+\frac{x^8}{8!}\right)e^{3x}$$

3.5.11. The *Bell number* $B(n)$ denotes the number of ways to partition a set of n distinct objects into any number of nonempty subsets, with $B(0)=1$. Clearly $B(1)=1$, since $\{a\}$ itself is the only nonempty subset of itself; moreover $B(2)=2$, since $\{a, b\}$ can be partitioned in two ways: $\{a, b\}$ itself or the two singleton sets $\{a\}$ and $\{b\}$.

(a) Verify that $B(3)=5$ and $B(4)=15$.

(b) Use combinatorial reasoning to derive the recursion formula

$$B(n+1)=\sum_{r=0}^{n}\binom{n}{r}B(r)$$

(c) Let $B(x)=\sum_{n\geq 0}B(n)\frac{x^n}{n!}$ be the EGF of the Bell numbers. Use the result of Problem 3.5.5 to show that $B'(x)=e^xB(x)$.

(d) Solve the differential equation obtained in part (c) to show that $B(x)=e^{(e^x-1)}$

Answer

(a) The 5 partitions of $\{a, b, c\}$ are:

$$\{a, b, c\} \quad \{a\}, \{b, c\} \quad \{b\}, \{a, c\} \quad \{c\}, \{a, b\} \quad \{c\}, \{a\}, \{b\}$$

The 15 partitions of $\{a, b, c, d\}$ are:

$$\begin{array}{ccccc} \{a, b, c, d\} & \{a\}, \{b, c, d\} & \{b\}, \{a, c, d\} & \{c\}, \{a, b, d\} & \{d\}, \{a, b, c\} \\ \{a, b\}, \{c, d\} & \{a, c\}, \{b, d\} & \{a, d\}, \{b, c\} & \{a\}, \{b\}, \{c, d\} & \{a\}, \{c\}, \{b, d\} \\ \{a\}, \{d\}, \{b, c\} & \{b\}, \{c\}, \{a, d\} & \{b\}, \{d\}, \{a, c\} & \{c\}, \{d\}, \{a, b\} & \{a\}, \{b\}, \{c\}, \{d\} \end{array}$$

(b) Suppose $[n + 1]$ is the set distinct elements. Let r be the number of elements *not* in the same subset with $n + 1$, where $r = 0, 1, \ldots, n$. These r elements can be selected in $\binom{n}{r}$ ways, and partitioned into nonempty subsets in $B(r)$ ways. Summing over all r gives all of the ways to partition $n + 1$ distinct elements into nonempty subsets.

(c) The EGF of the sequence $B(n + 1)$ is $B'(x)$ by Problem 3.5.5. By Theorem 3.34, $\sum_{r=0}^{n} \binom{n}{r} B(r)$ is the coefficient of $\frac{x^n}{n!}$ of the product of the EGFs of the sequences $B(0), B(1), B(2), \ldots, B(n), \ldots$ and $1, 1, 1, \ldots, 1, \ldots$. Since these EGFs are $B(x)$ and e^x, respectively, using part (b) it follows that $B'(x) = e^x B(x)$.

(d) Since $\frac{d}{dx} \log B(x) = e^x$, we see that $\log B(x) = e^x + C$ for some constant of integration C. Moreover, $B(0) = 1$, so we see that $C = \log B(0) - e^0 = 0 - 1 = -1$. Therefore, $\log B(x) = e^x - 1$, which shows that $B(x) = e^{(e^x - 1)}$.

3.5.13. The product of the first n even positive integers is the "double" factorial $(2n)!! = (2n) \cdot (2n - 2) \cdot \cdots \cdot 4 \cdot 2$.

(a) Verify that $(2n)!! = 2^n n!$, where $0!! = 1$ by definition.

(b) Show that the EGF of the double factorial sequence of even positive integers is $\frac{1}{1 - 2x} = \sum_{n=0}^{\infty} (2n)!! \frac{x^n}{n!}$.

Answer

(a) $(2n)!! = (2n)(2n - 2) \cdot \cdots \cdot (2) = 2^n n!$

(b) $\sum_{n \geq 0} (2n)!! \frac{x^n}{n!} = \sum_{n \geq 0} 2^n n! \frac{x^n}{n!} = \sum_{n \geq 0} (2x)^n = \frac{1}{1 - 2x}$

3.5.15. Use combinatorial reasoning to prove that

$P(j+k,n) = \sum_{r=0}^{n} \binom{n}{r} P(j,r)P(k,n-r)$ where j and k are nonnegative integers.

Answer

Suppose that j women and k men belong to the Combinatorics Club. How many ways can n members of the club line up for a photo?

Answer 1. There are $P(j+k,n)$ ways to line up.

Answer 2. Suppose r women are to be in the picture, where $r = 0, 1, 2, \dots, n$. There are $\binom{n}{r}$ ways to select the n positions to be taken by a woman, and $P(j,r) = (j)_r$ ways to choose an r-permutations of women to take these positions. The remaining $n - r$ remaining positions can be occupied by men in $P(k, n-r) = (k)_{n-r}$ ways. Altogether, there are $\binom{n}{r}(j)_r(k)_{n-r}$ ways to line up with r women. Summing over r gives the identity.

3.5.17. The $n!$ permutations of an n-set contain $\binom{n}{k} D_{n-k}$ permutations that have exactly k fixed points, since there are $\binom{n}{k}$ ways to choose a subset of k elements that are fixed and D_{n-k} ways to derange the remaining $n - k$ elements. The average number μ_n of fixed points of the $n!$ permutations is then given by

$$\mu_n = \frac{1}{n!}\sum_{k=0}^{n} k\binom{n}{k}D_{n-k}$$

and the standard deviation σ_n of the number of fixed points is given by

$$\sigma_n^2 = \frac{1}{n!}\sum_{k=0}^{n}(k-\mu_n)^2\binom{n}{k}D_{n-k}$$

(a) Prove that the average number of fixed points is $\mu_n = 1$, all $n \geq 1$.

(b) Prove that the standard deviation of the fixed points of the permutations of an n-set is $\sigma_1 = 0$ and $\sigma_n = 1$ for all $n \geq 2$.

[*Hint*: Use Theorem 3.34 and the results of Problem 3.5.12.]

Answer

(a) Theorem 3.34 and the result of part (a) of Problem 3.5.12 show that

$$\mu_n = [x^n]\left(xe^x D(x)\right) = [x^n]\left(xe^x \frac{e^{-x}}{1-x}\right)$$
$$= [x^n]\left(\frac{x}{1-x}\right) = [x^n]\left(x + x^2 + x^3 + \cdots\right) = 1, n \geq 1.$$

(An alternate solution was given in Problem 2.6.13.)

(b) Theorem 3.34 and the result of part (c) of problem 3.5.12 show that

$$\sigma_n^2 = [x^n]\left(\left(x^2 - x + 1\right) e^x D(x)\right) = [x^n]\left(\left(x^2 - x + 1\right) e^x \frac{e^{-x}}{1-x}\right)$$
$$= [x^n]\left(\frac{x^2}{1-x} + 1\right) = [x^n]\left(1 + x^2 + x^3 + x^4 + \cdots\right) = \begin{cases} 0, \text{if } n = 1 \\ 1, \text{if } n \geq 2 \end{cases}.$$

3.5.19. In how many ways can a 10-letter word be formed from the letters A, B, C, D, and E, where each letter must appear at least once?

Answer

Each of the five letters has the same EGF, $f^{(e)}(x) = e^x - 1$, so the number of 10 letter "words" is

$$\left[\frac{x^{10}}{10!}\right]\left(e^x - 1\right)^5 = \left[\frac{x^{10}}{10!}\right]\left(e^{5x} - 5e^{4x} + 10e^{3x} - 10e^{2x} + 5e^x - 1\right)$$
$$= 5^{10} - 5 \cdot 4^{10} + 10 \cdot 3^{10} - 10 \cdot 2^{10} + 5.$$

3.5.21. In how many ways can n pieces of fruit be lined up on a shelf, where there must be an even number of both apples and kiwi fruits and an odd number of both bananas and oranges?

Answer

The numbers of apples and kiwis both have the EGF $\cosh x$ and the numbers of bananas and oranges each have the EGF $\sinh x$. The product of the four EGFs is

$$\cosh^2 x \sinh^2 x = (\cosh x \sinh x)^2 = \left(\frac{1}{2}\sinh(2x)\right)^2$$
$$= \frac{\sinh^2(2x)}{4} = \frac{\cosh(4x) - 1}{8}.$$

Therefore, the number of ways to line up the n pieces of fruit is

$$\left[\frac{x^n}{n!}\right]\frac{\cosh(4x) - 1}{8} = \begin{cases} 2 \cdot 4^{n-2} & , n \geq 2 \text{ and even} \\ 0 & , n = 0 \text{ or odd.} \end{cases}$$

Alternatively, using just exponentials,

$$\left[\frac{x^n}{n!}\right]\left(\frac{e^x - e^{-x}}{2}\right)^2\left(\frac{e^x + e^{-x}}{2}\right)^2 = \left[\frac{x^n}{n!}\right]\left(\frac{e^{2x} - e^{-2x}}{4}\right)^2$$
$$= \left[\frac{x^n}{n!}\right]\left(\frac{e^{4x} - 2 + e^{-4x}}{16}\right)$$
$$= \begin{cases} 2 \cdot 4^{n-2} & , n \geq 2 \text{ and even} \\ 0 & , n = 0 \text{ or odd.} \end{cases}$$

3.5.23. Let c_n be the number of ways to seat n people around any number of indistinguishable circular tables, with $c_0 = 1$. For example, $c_1 = 1$ and $c_2 = 2$.

(a) Verify that $c_3 = 6$.

(b) Verify the recursion formula

$$c_{n+1} = \sum_{r=0}^{n} \binom{n}{r} c_r (n-r)!$$

(c) Let $C(x)$ be the EGF of the sequence $c_0, c_1, c_2, \ldots, c_n, \ldots$. Use the result of Problem 3.5.5 and Theorem 3.34 to show that $C'(x) = \dfrac{C(x)}{1-x}$.

(d) Use the result of part (c) to find $C(x)$ and give a general formula for c_n.

(e) You may be surprised at your answer for part (c). Use combinatorial reasoning to give a more direct derivation by calculating in how many ways can person 1 be seated; person 2 be seated once person 1 has been seated; … ; person $r + 1$ be seated once persons 1 through r have been seated.

Answer

(a) There are $(3 - 1)! = 2$ ways to seat all three at the same table, 3 ways to seat two at a table and one at another table, and 1 way to seat the three people each alone at a table.

(b) There are $\binom{n}{r} c_r$ ways to choose and seat r persons at tables not shared with person $n + 1$. This leaves $n - r$ people to share the table with person $n + 1$, and they can be seated in $(n - r)!$ ways. Summing over $r = 0, 1, \ldots, n$ gives the c_{n+1} ways to seat all $n + 1$ people at indistinguishable circular tables.

(c) The EGF of the sequence c_{n+1} is $C'(x)$ by problem 3.5.5. By Theorem 3.34, $\sum_{r=0}^{n}\binom{n}{r}c_r\,(n-r)!$ is the coefficient of $\frac{x^n}{n!}$ of the product of the EGFs of the sequences $c_0, c_1, c_2, \ldots, c_n, \ldots$ and $0!, 1!, 2!, \ldots, s!, \ldots$. Since these EGFs are $C(x)$ and $\frac{1}{1-x}$, respectively, it follows that $C'(x) = \frac{C(x)}{1-x}$.

(d) The differential equation for $C(x)$ found in part (c) can be rewritten as $\frac{d}{dx}\log C(x) = \frac{1}{1-x} = \frac{d}{dx}\log\frac{1}{1-x}$, so $C(x) = \frac{K}{1-x}$ for some constant K. Since $C(0) = c_0 = 1$, we see that $C(x) = \frac{1}{1-x}$. Thus $c_n = \left[\frac{x^n}{n!}\right]\frac{1}{1-x} = \left[\frac{x^n}{n!}\right]\sum_{n\geq 0}x^n = \left[\frac{x^n}{n!}\right]\sum_{n\geq 0}n!\frac{x^n}{n!} = n!$.

(e) There is just one way to seat person 1 at a circular table, so $c_1 = 1$. Person 2 can either sit immediately to the right of person 1 or at a new table, so $c_2 = 2$. If persons 1 through r have been seated in any of the possible c_r ways at circular tables, person $r+1$ can either be seated to the immediate right of a person already seated in r ways, or else be seated alone at a table. Thus, person $r+1$ can be seated in $(r+1)\,c_r$ ways, giving us the recursion $c_{r+1} = (r+1)\,c_r$. By iteration (or mathematical induction) it follows that $c_n = n!$.

3.5.25. A group of adventurers decides to split into two groups. Group A will ride the two-leg zip line, where riders on the first leg are in the order of decreasing age and the second leg is arranged so no one is in the same position as on the first leg. The second group B will take a swim in Lago Crocodilo.

(a) How many ways can the n adventurers pursue the activities?

(b) Describe the ways for $n = 3$ adventurers to pursue the activities.

Answer

(a) If r people ride the zip line, there are $a_r = D_r$ ways to rearrange themselves for the second leg, where D_r is the rth derangement number. The EGF for the zip line is therefore $f_A^{(e)}(x) = D(x) = \frac{e^{-x}}{1-x}$. There is just one way, $b_s = 1$, for s people to take the swim, so the EGF for swimming is $f_B^{(e)}(x) = \sum_{s\geq 0}\frac{x^s}{s!} = e^x$. The number of ways to pursue the two activities is therefore

$$\left[\frac{x^n}{n!}\right]D(x)\,e^x = \left[\frac{x^n}{n!}\right]\frac{e^{-x}e^x}{1-x} = \left[\frac{x^n}{n!}\right]\frac{1}{1-x} = n!\,\left[x^n\right]\sum_{n\geq 0}x^n = n!.$$

(b) There are $D_3 = 2$ ways all three can ride the zip line. There are 3 ways one can choose the swim and the two others ride the zip line. There is 1 way all three can do the swim. Altogether, there are $2 + 3 + 1 = 6 = 3!$ ways for 3 adventurers to pursue the activities, in agreement with part (a).

3.5.27. An unlimited supply of red and white tiles is available.

(a) In how many ways can 10 tiles be selected if there must be an odd number of each color?

(b) In how many ways can a 1×10 board be tiled with an odd number of red and an odd number of white tiles?

Answer

(a) This is a combination problem, so OGFs are appropriate. $[x^{10}]\left(x + x^3 + x^5 + \cdots\right)^2 = [x^{10}]\dfrac{x^2}{\left(1 - x^2\right)^2} = [x^{10}]\sum_{n=0}^{\infty} nx^{2n} = 5$. Of course, the answer is obvious without OGFs: 1, 3, 5, 7, or 9 red tiles are selected, together with 9, 7, 5, 3, or 1 respective white tiles.

(b) This is a permutation problem, so EGFs are appropriate. $\left[\dfrac{x^{10}}{10!}\right](\sinh x)^2 = \left[\dfrac{x^{10}}{10!}\right]\dfrac{\cosh 2x - 1}{2} = \dfrac{2^{10}}{2} = 2^9 = 512$. Alternatively suppose that r red and $10 - r$ white tiles are selected and arranged, where $r = 1, 3, 5, 7, 9$.. They can be arranged in $\binom{10}{r}$ ways, so there are at total of $\binom{10}{1} + \binom{10}{3} + \binom{10}{5} + \binom{10}{7} + \binom{10}{9} = 512$ arrangements.

PROBLEM SET 3.6

3.6.1. Use the geometric series $\dfrac{1}{1-x} = \sum_{n=0}^{\infty} x^n$ to calculate

(a) $[x^n]\dfrac{1}{1-5x}$ **(b)** $[x^n]\dfrac{1}{1-x^3}$ **(c)** $[x^n]\dfrac{1}{1+x}$ **(d)** $[x^{50}]\dfrac{x^7}{1+2x}$

Answer

(a) 5^n

(b) 1 if $n = 3k$, 0 otherwise

(c) 1 if n even, and -1 if n odd

(d) $[x^{50}]\, x^7 \sum_{n=0}^{\infty}(-2x)^n = [x^{43}]\sum_{n=0}^{\infty}(-2x)^n = (-2)^{43} = -2^{43}$

3.6.3. Let $\varphi = \frac{1+\sqrt{5}}{2}$ and $\hat{\varphi} = \frac{1-\sqrt{5}}{2}$ be the two roots of the polynomial equation $x^2 = x + 1$. Then $\varphi^2 = \varphi + 1$ so that $\varphi^n = \varphi^{n-1} + \varphi^{n-2}$ and it is seen that $a_n = \varphi^n$ is a sequence that satisfies the Fibonacci recursion relation $a_n = a_{n-1} + a_{n-2}, n \geq 2$, with the initial values $a_0 = 1, a_1 = \varphi$. Similarly, the sequence $b_n = \hat{\varphi}^n$ satisfies $b_0 = 1, b_1 = \hat{\varphi}, b_n = b_{n-1} + b_{n-2}, n \geq 2$. Now show that any generalized Fibonacci sequence $c_0, c_1, c_n = c_{n-1} + c_{n-2}, n \geq 2$, can be written in the form $c_n = r\varphi^n + s\hat{\varphi}^n$ if the constants r and s are chosen suitably.

Answer

It is evident that the sequence $r\varphi^n + s\hat{\varphi}^n$ satisfies the Fibonacci recursion relation, so it only remains to determine the coefficients r and s so that $c_0 = r\varphi^0 + s\hat{\varphi}^0 = r + s$ and $c_1 = r\varphi^1 + s\hat{\varphi}^1 = r\varphi + s\hat{\varphi}$. These two conditions are combined by the matrix equation $\begin{bmatrix} 1 & 1 \\ \varphi & \hat{\varphi} \end{bmatrix} \begin{bmatrix} r \\ s \end{bmatrix} = \begin{bmatrix} c_0 \\ c_1 \end{bmatrix}$. Since $\det \begin{bmatrix} 1 & 1 \\ \varphi & \hat{\varphi} \end{bmatrix} = \hat{\varphi} - \varphi = -\sqrt{5} \neq 0$, there is a unique solution $\begin{bmatrix} r \\ s \end{bmatrix} = \begin{bmatrix} 1 & 1 \\ \varphi & \hat{\varphi} \end{bmatrix}^{-1} \begin{bmatrix} c_0 \\ c_1 \end{bmatrix} = \frac{1}{\hat{\varphi} - \varphi} \begin{bmatrix} \hat{\varphi} & -1 \\ -\varphi & 1 \end{bmatrix} \begin{bmatrix} c_0 \\ c_1 \end{bmatrix}$. That is, $r = \frac{1}{\sqrt{5}} \left(c_1 - c_0\hat{\varphi}\right)$ and $s = \frac{1}{\sqrt{5}} \left(c_0\varphi - c_1\right)$.

3.6.5. Let $f_F(x)$ and $f_L(x)$ be the OGFs of the Fibonacci and Lucas numbers (see Problem 3.6.4).

(a) Show that $\frac{(2-x)\left(1+x^2\right)}{1-x-x^2} - 2 - x = \frac{5x^2}{1-x-x^2}$.

(b) Explain carefully why $L_{n+2} + L_n = 5F_{n+1}, n \geq 0$, using part (a).

Answer

(a)

$$\begin{aligned} \frac{(2-x)\left(1+x^2\right)}{1-x-x^2} - 2 - x &= \frac{2 - x + 2x^2 - x^3 - (2+x)\left(1-x-x^2\right)}{1-x-x^2} \\ &= \frac{2 - x + 2x^2 - x^3 - 2 + 2x + 2x^2 - x + x^2 + x^3}{1-x-x^2} \\ &= \frac{5x^2}{1-x-x^2} \end{aligned}$$

(b) Part (a) shows that $\left(1+x^2\right) f_L(x) - 2 - x = 5x f_F(x)$, so $L_{n+2} + L_n = \left[x^{n+2}\right]\left(\left(1+x^2\right) f_L(x) - 2 - x\right) = \left[x^{n+2}\right] 5x f_F(x) = \left[x^{n+1}\right] 5 f_F(x) = 5F_{n+1}$.

3.6.7. OGF of the Perrin Sequence Let ρ, σ, and τ be the roots of the cubic polynomial equation $x^3 = x + 1$, so that $x^3 - x - 1 = (x - \rho)(x - \sigma)(x - \tau)$. Derive the OGF of the Perrin sequence $p_0 = 3, p_1 = 0, p_2 = 2, p_n = p_{n-2} + p_{n-3}, n \geq 3$ by carrying out the following steps.

(a) Compare coefficients of the equation $x^3 - x - 1 = (x - \rho)(x - \sigma)(x - \tau)$ to show that $\rho + \sigma + \tau = 0, \rho\sigma + \sigma\tau + \rho\tau = -1, \rho\sigma\tau = 1$.

(b) Use part (a) to show that $\rho^2 + \sigma^2 + \tau^2 = 2$.

(c) Let $\widehat{p}_n = \rho^n + \sigma^n + \tau^n$. Show that $\widehat{p}_n = p_n$ for all $n \geq 0$.

(d) Show that

$$\sum_{n=0}^{\infty} p_n x^n = \frac{3 - x^2}{1 - x^2 - x^3}$$

Answer

(a) $x^3 - x - 1 = (x - \rho)(x - \sigma)(x - \tau) = x^3 - (\rho + \sigma + \tau)x^2 + (\rho\sigma + \sigma\tau + \rho\tau)x - \rho\sigma\tau$.

(b) $0 = (\rho + \sigma + \tau)^2 = \rho^2 + \sigma^2 + \tau^2 + 2\rho\sigma + 2\sigma\tau + 2\rho\tau = \rho^2 + \sigma^2 + \tau^2 - 2$.

(c) Since $\rho^3 = \rho + 1, \sigma^3 = \sigma + 1$, and $\tau^3 = \tau + 1$, we see that $\rho^n = \rho^{n-2} + \rho^{n-3}, \sigma^n = \sigma^{n-2} + \sigma^{n-3}$, and $\tau^n = \tau^{n-2} + \tau^{n-3}$. Therefore, $\widehat{p}_n = \rho^n + \sigma^n + \tau^n = \left(\rho^{n-2} + \rho^{n-3}\right) + \left(\sigma^{n-2} + \sigma^{n-3}\right) + \left(\tau^{n-2} + \tau^{n-3}\right) = \widehat{p}_{n-2} + \widehat{p}_{n-3}$. That is, $\widehat{p}_n = \rho^n + \sigma^n + \tau^n$ satisfies the Perrin recursion relation. Therefore, it is enough to show that both $\hat{p}_n$and p_n have the same values for $n = 0$, 1, and 2. This is easily verified: $\widehat{p}_0 = \rho^0 + \sigma^0 + \tau^0 = 3, \widehat{p}_1 = \rho + \sigma + \tau = 0, \widehat{p}_2 = \rho^2 + \sigma^2 + \tau^2 = 2$. so the initial three values $\widehat{p}_0 = p_0 = 3, \widehat{p}_1 = p_1 = 0, \widehat{p}_2 = p_2 = 2$ agree. Therefore, $\widehat{p}_n = p_n = \rho^n + \sigma^n + \tau^n, n \geq 0$.

(d)

$$\sum_{n=0}^{\infty} p_n x^n = \sum_{n=0}^{\infty} (\rho^n + \sigma^n + \tau^n)x^n = \frac{1}{1 - \rho x} + \frac{1}{1 - \sigma x} + \frac{1}{1 - \tau x}$$

$$= \frac{\left(1 - (\sigma + \tau)x + \sigma\tau x^2\right) + \left(1 - (\rho + \tau)x + \rho\tau x^2\right) + \left(1 - (\rho + \sigma)x + \rho\sigma x^2\right)}{(1 - \rho x)(1 - \sigma x)(1 - \tau x)}$$

$$= \frac{3 - 2(\rho + \sigma + \tau)x + (\sigma\tau + \rho\tau + \rho\sigma)x^2}{1 - (\rho + \sigma + \tau)x + (\sigma\tau + \rho\tau + \rho\sigma)x^2 - \rho\sigma\tau x^3} = \frac{3 - x^2}{1 - x^2 - x^3}$$

3.6.9. Ana discovered six boxes of stepping stones in her basement. Each box contains five identical stones, and each box contains a different color of stone. How many ways can she make a path with

(a) all 30 stones? **(b)** 29 stones? **(c)** 15 stones?

(Evaluate numerically if a CAS is available.)

Answer

(a) $\binom{30}{5,5,5,5,5,5} = \frac{30!}{(5!)^6} = 88,832,646,059,788,350,720$

(b) $6\binom{29}{4,5,5,5,5,5} = 6\frac{29!}{4!\,(5!)^5} = 88,832,646,059,788,350,720$

(c) $\left[\frac{x^{15}}{15!}\right]\left(1+\frac{x}{1!}+\frac{x^2}{2!}+\frac{x^3}{3!}+\frac{x^4}{4!}+\frac{x^5}{5!}\right)^6 = 393,402,129,120$

4

ALTERNATING SUMS, INCLUSION-EXCLUSION PRINCIPLE, ROOK POLYNOMIALS, AND FIBONACCI NIM

PROBLEM SET 4.2

4.2.1. An *involution* is a bijection $f : X \to X$ for which $f(f(x)) = x$ for all $x \in X$. Show that each of the following functions is an involution.

(a) $f(x) = -x, X = \mathbb{R}$ ($\mathbb{R}$ is the set of real numbers)

(b) $g(x) = \dfrac{1}{x}, X = \mathbb{R} - \{0\}$

(c) $h(x) = -\dfrac{1}{x},\ X = \mathbb{R} - \{0\}$

(d) $k(x) = a + \dfrac{1}{x-a}, X = \mathbb{R} - \{a\}, a \in \mathbb{R}$

Answer

(a) $f(f(x)) = f(-x) = -(-x) = x$

(b) $g(g(x)) = g\left(\dfrac{1}{x}\right) = \dfrac{1}{\left(\dfrac{1}{x}\right)} = x$

(c) $h(h(x)) = h\left(-\dfrac{1}{x}\right) = -\dfrac{1}{\left(-\dfrac{1}{x}\right)} = x$

Solutions Manual to Accompany Combinatorial Reasoning: An Introduction to the Art of Counting, First Edition. Duane DeTemple and William Webb.

(d) $k(k(x)) = a + \dfrac{1}{k(x) - a} = a + \dfrac{1}{a + \dfrac{1}{x-a} - a} = a + (x - a) = x$

4.2.3. Prove that

$$\sum_{k=0}^{m} (-1)^k \binom{n}{k} = (-1)^m \binom{n-1}{m}$$

for all $m \geq 0$ and $n > 0$.

Answer

The formula is certainly correct for $m \geq n$, since then $\sum_{k=0}^{m} (-1)^k \binom{n}{k} = \sum_{k=0}^{n} (-1)^k \binom{n}{k} = 0$ and $\binom{n-1}{m} = 0$. In the case that $m < n$, the DIE method can be used to evaluate the alternating sum $\sum_{k=0}^{m} (-1)^k \binom{n}{k}$:

D. For any $k = 0, 1, 2, \ldots, m$, there are $\binom{n}{k}$ subsets of $[n]$ with k elements.

I. Pair any k-element subset to either a $k + 1$ element subset by appending the element 1, or pair it to a $k - 1$ element subset by deleting the element 1 which is sign reversing.

E. The only unpaired subsets are those with m elements, none of which is the element 1.

There are $\binom{n-1}{m}$ unpaired subsets, each of sign $(-1)^m$, giving us the desired sum.

4.2.5. **(a)** Why is

$$\sum_{k=0}^{n} (-1)^k \binom{n}{k} k^m = 0$$

for all $m < n$?

(b) Why is $\sum_{k=0}^{n} (-1)^k \binom{n}{k} k^n = (-1)^n n!$ for all $n > 0$?

Answer

(a) By Example 4.7, the sum is $(-1)^n T(m, n)$, which is zero for $m < n$.

(b) By Example 4.7, the sum is $(-1)^n T(n, n) = (-1)^n n!$.

4.2.7. Prove that

$$\sum_{k=0}^{m} (-1)^k \binom{m}{k} \binom{n-k}{r-k} = \binom{n-m}{r}, \quad m + r \leq n$$

[*Hint*: Consider tilings of a $1 \times n$ board with red, blue, and white tiles, where a total of r tiles are red or blue and the red tiles are allowed only on a subset of the first m cells of the board.]

Answer

D. $\sum_{k=0}^{m} \binom{m}{k} \binom{n-k}{r-k}$ counts the number of ways to tile a $1 \times n$ board with red, blue and white tiles, where there are a total of r red and blue tiles, and the red tiles are only allowed to be placed on a subset of the first m cells of the $1 \times n$ board. Here k represents the number of red tiles used, for $k = 0, 1, \ldots, m$.

I. The leftmost red or blue tile can be switched in color from red to blue or vice versa, which is sign reversing.

E. The exceptional tilings are those with white tiles only covering the first m cells of the board: there are then r blue tiles among the last $n - m$ cells of the board.

There are $\binom{n-m}{r}$ exceptional tilings, all of sign $(-1)^0 = +1$.

4.2.9. Show that $(-1)^n \sum_{k=0}^{n} (-1)^k \binom{n}{k} k^{n+1} = n\frac{(n+1)!}{2}$ by

(a) using Example 4.7

(b) direct application of the DIE method

Answer

(a) By Example 4.7, the sum is

$$T(n+1, n) = \binom{n+1}{2} T(n, n) = \frac{(n+1)n}{2} n! = \frac{n(n+1)!}{2}.$$

(b) Consider the sum $\sum_{k=0}^{n}(-1)^k\binom{n}{k}k^{n+1}$.

D. $\binom{n}{k}k^{n+1}$ is the number of ways to distribute $n+1$ distinct objects to a subset of k distinct recipients chosen from a numbered group of n recipients.

I. Let A be the subset of $n-k$ not chosen and B be the set that were chosen to potentially be a recipient but were not assigned any object by the distribution. The smallest member of A or B can be moved to the opposite set when at least one of A or B is not empty, which is sign reversing.

E. The exceptional distributions occur when A and B are both empty. That is, all n members of the group are assigned at least one of the $n+1$ objects. There are n ways to choose the person to hold out both hands to get an object, with the others holding out just their right hands. The $n+1$ objects can be placed in people's hands in $(n+1)!$ ways. The person with an object in each hand can switch hands and still get the same objects, so altogether there are $\frac{n\,(n+1)!}{2}$ exceptional distributions each of sign $(-1)^n$.

We now see that $\sum_{k=0}^{n}(-1)^k\binom{n}{k}k^{n+1}=(-1)^n\frac{n\,(n+1)!}{2}$, which is equivalent to the formula to be proved.

4.2.11. Prove that

$$\sum_{k=m}^{n-1}(-1)^k\binom{n}{k+1}\binom{k}{m}=(-1)^m$$

for all integers $m, n \geq 0$.

[*Hint*: Consider the committee-subcommittee selection model in a club of n members that are numbered $1, 2, \ldots, n$. Let the committee be chaired by the committee member with the highest number. Consider the set A of the club members not on the committee with numbers lower than the chair, and the set B of committee members not on the subcommittee and not the chair.]

Answer

D. $\binom{n}{k+1}$ is the number of ways to choose a committee of $k+1$ members from a club of n members. Assuming the members are numbered 1 through n, let the committee member with the highest number be the chair, and then choose a subcommittee of size m from the k members on the committee that are not the chair, in $\binom{k}{m}$ ways.

I. Consider these two sets: the set A of club members not on the committee with a lower number than the chair, and the set B of members on the committee that are not the chair nor are on the subcommittee. The lowest member of $A \cup B$ can be switched to the opposite set. Since this either reduces or increases the committee membership by 1, it is a sign-reversing involution.

E. The only unpaired committee and subcommittee selection occurs when $A = B = \varnothing$. That is, the committee consists of the members $1, 2, \ldots, m+1$ chaired by $m+1$, and the subcommittee are the club members $1, 2, \ldots, m$.

Since the algebraic sign of one exception is $(-1)^m$, we see this evaluates the alternating sum.

4.2.13. Use equation (4.13) to explain why

$$\sum_{k\geq 0} (-1)^{n-k} \binom{n-k}{k} = \begin{cases} 1, & \text{if } n \equiv 0 \pmod 3 \\ 0, & \text{if } n \equiv 2 \pmod 3 \\ -1, & \text{if } n \equiv 1 \pmod 3 \end{cases}$$

Answer

Denoting the sum by t_n, it follows that $t_n = (-1)^n s_n$ by equations (4.13) and (4.14). But then $t_n = (-1)^n s_n = (-1)^n \left(-s_{n-3}\right) = (-1)^{n-3} s_{n-3} = t_{n-3}$. Since $t_1 = -s_1 = -1, t_2 = s_2 = 0, t_3 = -s_1 = 1$, the recursion $t_n = t_{n-3}$ proves the given formula.

4.2.15. In Theorem 4.5, it was shown that

$$\sum_{k\geq 0} (-1)^k \binom{n}{k} \binom{k}{m} = (-1)^n \delta_{n,m}$$

Use this identity to prove that

$$a_n = \sum_{i\geq 0} (-1)^i \binom{n}{i} b_i \text{ if and only if } b_n = \sum_{j\geq 0} (-1)^j \binom{n}{j} a_j.$$

Answer

Suppose that $a_k = \sum_{j\geq 0} (-1)^i \binom{k}{j} b_j$. Then

$$\sum_{k\geq 0} (-1)^k \binom{n}{k} a_k = \sum_{k\geq 0} (-1)^k \binom{n}{k} \left(\sum_{j\geq 0} (-1)^j \binom{k}{j} b_j \right)$$

$$= \sum_{j\geq 0} (-1)^j \left(\sum_{k\geq 0} (-1)^k \binom{n}{k} \binom{k}{j} \right) b_j$$

$$= \sum_{j\geq 0} (-1)^j (-1)^n \delta_{n,j} b_j = b_n.$$

The converse holds by the symmetry of the two formulas.

4.2.17. The $n!$ permutations of $[n]$ can be partitioned according to the exact number k of elements that are deranged, leaving the remaining $n - k$ elements fixed. Therefore, $n! = \sum_{k\geq 0} \binom{n}{k} D_k$. Use the result shown in Problem 4.2.15 to obtain formula (4.10) for D_n.

Answer

Let $a_n = n!$ and $b_n = (-1)^n D_n$. Then

$$a_n = n! = \sum_{k\geq 0} (-1)^k \binom{n}{k} (-1)^k D_k = \sum_{k\geq 0} (-1)^k \binom{n}{k} b_k$$

so by Problem 4.2.15,

$$D_n = (-1)^n b_n = (-1)^n \sum_{k\geq 0} (-1)^k \binom{n}{k} a_k$$

$$= (-1)^n \sum_{k\geq 0} (-1)^k \binom{n}{k} k! = \sum_{k=0}^{n} (-1)^{n-k} \frac{n!}{(n-k)!k!} k! = n! \sum_{j=0}^{n} \frac{(-1)^j}{j!}.$$

PROBLEM SET 4.3

4.3.1. How many of the first one million-positive integers are

(a) either a square or a cube?

(b) neither a square nor a cube?

Answer

(a) Let $U = [10^6]$, and consider the subsets

$$A_1 = \{1^2, 2^2, 3^2, \dots, 1000^2\}, \text{ and } A_2 = \{1^3, 2^3, 3^3, \dots, 100^3\}.$$

Then $A_1 \cap A_2 = \{1^6, 2^6, \dots, 10^6\}$, so the number of squares and cubes is $1000 + 100 - 10 = 1090$.

(b) $1{,}000{,}000 - 1090 = 998{,}910$.

4.3.3. How many of the first 1000 positive integers are divisible by either 6, 7, or 8?

Answer

Let A_1, A_2, and A_3 be, respectively, the sets of integers no larger than 1000 that are divisible by 6, 7, and 8. Since

$$\left\lfloor \frac{1000}{6} \right\rfloor = 166, \left\lfloor \frac{1000}{7} \right\rfloor = 142, \text{and } \left\lfloor \frac{1000}{8} \right\rfloor = 125,$$

it follows that $|A_1| = 166$, $|A_2| = 142$, and $|A_3| = 125$. Any integer divisible by 6 and 7, or by 7 and 8, is divisible by 42 and 56 respectively. Therefore $|A_1 \cap A_2| = 23$ and $|A_2 \cap A_3| = 17$, since $\left\lfloor \frac{1000}{42} \right\rfloor = 23$ and $\left\lfloor \frac{1000}{56} \right\rfloor = 17$. The least common multiple of 6 and 8 is 24 and $\left\lfloor \frac{1000}{24} \right\rfloor = 41$, so $|A_1 \cap A_3| = 41$. Finally, the least common multiple of 6, 7, and 8 is 168 and $\left\lfloor \frac{1000}{168} \right\rfloor = 5$, so $|A_1 \cap A_2 \cap A_3| = 5$. Therefore $(166 + 142 + 125) - (23 + 17 + 41) + 5 = 357$ of the positive integers not larger than 1000 are divisible by at least one of 6, 7, or 8.

4.3.5. How many permutations of the multiset $S = \{3{\cdot}X, 4{\cdot}Y, 2{\cdot}Z\}$ contain none of the blocks XXX, YYYY, ZZ?

Answer

Let U be the set of unrestricted permutations of S, so $|U| = \binom{9}{3, 4, 2} = \frac{9!}{3!4!2!}$. Let A_1, A_2, and A_3 be, respectively, the sets of permutations that contain the substring XXX, YYYY, or ZZ. Therefore,

$$|A_1| = \frac{7!}{1!4!2!}, |A_2| = \frac{6!}{3!1!2!}, |A_3| = \frac{8!}{3!4!1!}, |A_1 \cap A_2| = \frac{4!}{1!1!2!},$$

$$|A_1 \cap A_3| = \frac{6!}{1!4!1!}, |A_2 \cap A_3| = \frac{5!}{3!1!1!}, \text{and } |A_1 \cap A_2 \cap A_3| = 3!.$$

By PIE, there are

$$\left|\overline{A_1} \cap \overline{A_2} \cap \overline{A_3}\right| = \frac{9!}{3!4!2!} - \left(\frac{7!}{4!2!} + \frac{6!}{3!2!} + \frac{8!}{3!4!}\right) + \left(\frac{4!}{2!} + \frac{6!}{4!} + \frac{5!}{3!}\right) - 3!$$

$$= 1260 - (105 + 60 + 280) + (12 + 30 + 20) - 6 = 871$$

permutations not containing the forbidden blocks.

4.3.7. In a poll of 100 voters, asking them if they had a favorable opinion about candidates A, B, and C, 40 liked candidate A, 47 liked candidate B, 53 liked candidate C, 7 liked both A and B, 28 liked B and C, and 13 liked A and C. Another three liked none of the candidates. How many of those polled liked all of the candidates?

Answer
Rearranging the PIE formula shows that

$$\begin{aligned}|A \cap B \cap C| &= |U| - (|A| + |B| + |C|) + (|A \cap B| + |B \cap C| + |A \cap C|) \\ &\quad - \left|\overline{A} \cap \overline{B} \cap \overline{C}\right| \\ &= 100 - (40 + 47 + 53) + (7 + 28 + 13) - 3 = 5.\end{aligned}$$

4.3.9. How many permutations of the letters in WHATSUPDOC do not contain any of the substrings WHAT, UP, or DOC?

Answer
Let A_1, A_2, and A_3 be, respectively, the sets of permutations that contain the substring WHAT, UP, or DOC. Thus,

$$|A_1| = 7!, |A_2| = 9!, |A_3| = 8!, |A_1 \cap A_2| = 6!,$$

$$|A_1 \cap A_3| = 5!, |A_2 \cap A_3| = 7!, \text{and } |A_1 \cap A_2 \cap A_3| = 4!$$

Therefore, $10! - (7! + 9! + 8!) + (6! + 5! + 7!) - 4! = 3{,}226{,}416$ permutations do not contain any of the forbidden substrings.

4.3.11. **(a)** Use PIE to calculate the number of ways that a 5-card hand can be dealt from a 52-card deck with at least one card of every suit.

(b) Check your answer to part (a) by using a more direct approach.

Answer

(a) Let the suits be numbered 1, 2, 3, and 4, and let A_i be the 5-card combinations missing suit i. Thus

$$|A_i| = \binom{39}{5}, \left|A_i \cap A_j\right| = \binom{26}{5},$$

$$\left|A_i \cap A_j \cap A_k\right| = \binom{13}{5}, |A_1 \cap A_2 \cap A_3 \cap A_4| = 0.$$

Using PIE, there are $\binom{52}{5} - 4\binom{39}{5} + 6\binom{26}{5} - 4\binom{13}{5} + 0 =$ $685,464$ hands with no suit missing.

(b) Choose which suit has two cards, and then the ways cards can be chosen from the four suits: $4\binom{13}{2}\binom{13}{1}^3 = 685{,}464$.

4.3.13. **(a)** Use PIE to show there are six solutions in integers of the equation $t_1 + t_2 + t_3 = 16$, where $2 \le t_1 \le 6, -1 \le t_2 \le 4$, and $0 \le t_3 \le 8$.

(b) List the six solutions.

Answer

(a) First, make the change of variables $x_1 = t_1 - 2, x_2 = t_2 + 1, x_3 = t_3$, to obtain the equivalent equation

$$x_1 + x_2 + x_3 = 15, \text{ where } 0 \le x_1 \le 4, 0 \le x_2 \le 5, \text{and } 0 \le x_3 \le 8.$$

If U is the set of all solutions of $x_1 + x_2 + x_3 = 15$ in nonnegative integers, then $|U| = \left(\binom{3}{15}\right) = \binom{17}{15} = 136$. Now let A_1, A_2, and A_3 be, respectively, the sets of solutions of the $x_1 + x_2 + x_3 = 15$ for which $x_1 \ge 5, x_2 \ge 6, x_3 \ge 9$. Then

$$|A_1| = \left(\binom{3}{15-5}\right) = \binom{12}{10} = 66,$$

$$|A_2| = \left(\binom{3}{15-6}\right) = \binom{11}{9} = 55,$$

$$|A_3| = \left(\binom{3}{15-9}\right) = \binom{8}{6} = 28,$$

$$|A_1 \cap A_2| = \left(\!\!\binom{3}{15-5-6}\!\!\right) = \binom{6}{4} = 15,$$

$$|A_2 \cap A_3| = \left(\!\!\binom{3}{15-6-9}\!\!\right) = \binom{2}{0} = 1,$$

$$|A_1 \cap A_3| = \left(\!\!\binom{3}{15-5-9}\!\!\right) = \binom{3}{1} = 3, |A_1 \cap A_2 \cap A_3| = 0.$$

Using PIE, there are $\left|\overline{A_1} \cap \overline{A_2} \cap \overline{A_3}\right| = 136 - (66 + 55 + 28) + (15 + 1 + 3) - 0 = 6$ solutions.

(b) $(t_1, t_2, t_3) = (4,4,8), (5,4,7), (5,3,8), (6,4,6), (6,3,7), (6,2,8)$

4.3.15. What theorem is proved when PIE is applied to the n sets $A_1 = A_2 = \cdots = A_n = \{1\}$?

Answer

The union and all of the intersections have one element, so

$$1 = \sum_{1\le i\le n} 1 - \sum_{1\le i<j\le n} 1 + \sum_{1\le i<j<k\le n} 1 + \cdots + (-1)^{n+1} \cdot 1$$

$$= n - \binom{n}{2} + \binom{n}{3} - \binom{n}{4} + \cdots + (-1)^{n+1}$$

which shows that $\binom{n}{0} - \binom{n}{1} + \binom{n}{2} - \binom{n}{3} + \binom{n}{4} + \cdots + (-1)^n \binom{n}{n} = 0$. This is the result of equation (4.5).

4.3.17. The Euler-Fermat theorem states that if n and a have no common prime divisor, then $a^{\varphi(n)} - 1$ is divisible by n. Verify this theorem for:

(a) $n = 24$ and $a = 5$.

(b) $n = 17$ and $a = 2$

Answer

(a) $\varphi(24) = \varphi(2^3 3^1) = 24\left(1 - \frac{1}{2}\right)\left(1 - \frac{1}{3}\right) = 24\left(\frac{1}{2}\right)\left(\frac{2}{3}\right) = 8,$

$\frac{5^8 - 1}{24} = \frac{390{,}624}{24} = 16{,}276$

(b) $\varphi(17) = 17 - 1 = 16, \frac{2^{16} - 1}{17} = \frac{65{,}535}{17} = 3855$

PROBLEM SET 4.4

4.4.1. **(a)** In how many ways can four distinct rooks be placed on a 7×9 board?
(b) What if the rooks are identical?

Answer

(a) $(7 \cdot 9)(6 \cdot 8)(5 \cdot 7)(4 \cdot 6) = (7)_4(9)_4$
(b) $(7)_4(9)_4/4!$

4.4.3. Verify there are two rook placements for the 4×4 chess board in Figure 4.3, using rook polynomials.

Answer

$$R(x, \mathcal{B}) = 1 + 6x + 9x^2 + 4x^3$$
$$4! - 6 \cdot 3! + 9 \cdot 2! - 4 \cdot 1! = 24 - 36 + 18 - 4 = 2$$

4.4.5. Compute the rook polynomial $R(x, \mathcal{B})$ for each of these boards:
(a) a $1 \times n$ rectangle
(b) a $2 \times n$ rectangle
(c) a $3 \times n$ rectangle

Answer

(a) $1 + nx$
(b) $1 + 2nx + n(n-1)x^2$
(c) $1 + 3nx + 3n(n-1)x^2 + n(n-1)(n-2)x^3$

4.4.7. How many ways are there to place six rooks on this chessboard?

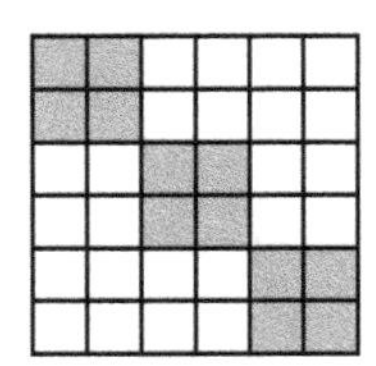

Answer
The rook polynomial is

$$\left(1 + 4x + 2x^2\right)^3 = 1 + 12x + 54x^2 + 112x^3 + 108x^4 + 48x^5 + 8x^6.$$

so $N = 6! - 12 \cdot 5! + 54 \cdot 4! - 112 \cdot 3! + 108 \cdot 2! - 48 \cdot 1! + 8 = 80$.

4.4.9. **(a)** Compute the rook polynomial $R(x, \mathcal{B})$ for this board by direct counting.

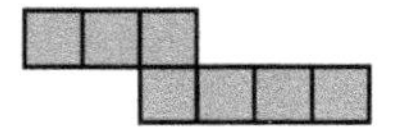

(b) Answer part (a) using equation (4.28), where $\boldsymbol{S}$ is the leftmost square of the second row.

Answer

(a) $1 + 7x + (2 \cdot 4 + 3)\, x^2 = 1 + 7x + 11x^2$

(b) $R(x, \mathcal{B}) = R(x, \mathcal{B} - \boldsymbol{S}) + xR\left(x, \mathcal{B}_{\boldsymbol{S}}\right) = (1 + 3x)^2 + x(1 + 2x) = 1 + 7x + 11x^2$

4.4.11. Obtain the rook polynomial of this board.

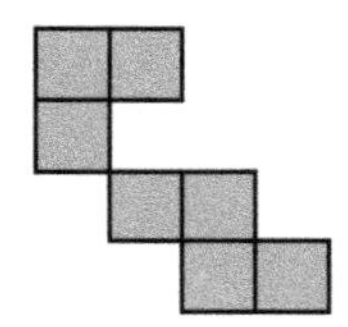

Answer

Using $\boldsymbol{S}$ as the left most square of row 3, $R(x, \mathcal{B}) = \left(1 + 3x + x^2\right)^2 + x(1 + 2x)^2 = 1 + 7x + 15x^2 + 10x^3 + x^4$

PROBLEM SET 4.5

4.5.1. Give the Zeckendorf representation of
(a) 27 **(b)** 37 **(c)** 86.

Answer

(a) $27 = 21 + 5 + 1$

(b) $37 = 34 + 3$

(c) $86 = 55 + 21 + 8 + 2$

4.5.3. Let $m = 101{,}010_Z$ and $n = 1{,}010{,}101_Z$. Without determining the base ten numerals of m and n, explain how to find the Zeckendorf numerals of
(a) $m + 1$ **(b)** $n + 1$ **(c)** $m + n$

Answer

Use the "carry" idea that … 0011 … = … 0100 … . Thus,

(a) $m + 1 = 101{,}010 + 1 = 101{,}011 = 101{,}100 = 110{,}000 = 1{,}000{,}000$

(b) $n + 1 = 1{,}010{,}101 + 1 = 1{,}010{,}110 = 1{,}011{,}000 = 1{,}100{,}000 = 10{,}000{,}000$

(c) $m + n = 101{,}010 + 1{,}010{,}101 = 1{,}111{,}111 = 10{,}011{,}111 = 10{,}100{,}111 = 10{,}101{,}001$

4.5.5. The stacked divisions by 5 shown below gives the sequence of remainders 2, 3, and 4.

$$\begin{array}{rl} 0 & R = 4 \\ 5\,\overline{)4} & R = 3 \\ 5\,\overline{)23} & R = 2 \\ 5\,\overline{)117} & \end{array}$$

(a) Explain why $117_{\text{ten}} = 4 \times 5^2 + 3 \times 5 + 2 = 432_{\text{five}}$.

In parts (b) and (c), use the stacked division method to determine the numerals that represent

(b) 1164_{ten} in base twelve, using the digits 0, 1, … , 9, T, E, where T is ten and E is eleven.

(c) 959_{ten} in base sixteen, with the digits 0, 1, … , 9, A, B, C, D, E, F.

Answer

(a) $117 = 23 \times 5 + 2 = (4 \times 5 + 3) \times 5 + 2 = 4 \times 5^2 + 3 \times 5 + 2$

(b) $1164_{\text{ten}} = 810_{\text{twelve}}$

$$\begin{array}{rl} 0 & R = 8 \\ 12\,\overline{)8} & R = 1 \\ 12\,\overline{)97} & R = 0 \\ 12\,\overline{)1164} & \end{array}$$

(c) $959_{\text{ten}} = 3\text{BF}_{\text{sixteen}}$

$$\begin{array}{rl} 0 & R = 3 \\ 16\,\overline{)3} & R = 11 = B \\ 16\,\overline{)59} & R = 15 = F \\ 16\,\overline{)959} & \end{array}$$

4.5.7. Recall that there are F_{n+2} binary sequences of length n for which no consecutive 1s appear (see Problem 1.5.11). Use this result to prove the uniqueness of the Zeckendorf representation.

Answer
The existence statement of Zeckendorf's theorem shows that every non-negative integer m from 0 to $F_{n+2} - 1$ corresponds to at least one binary string of length n with no consecutive 1s. But F_{n+2} is exactly the number of binary strings of length n with no consecutive 1s. Therefore, every m is represented by just one binary string with no consecutive 1s.

4.5.9. A greedy pirate and a smart pirate have discovered a treasure chest filled with gold bars weighing 25 pounds, 17 pounds, and 13 pounds. It is a long walk back through a crocodile infested swamp to the galleon, so the pirates know that a 60 pound load is the most each can carry. The greedy pirate put two of the heaviest bars in his pack. What did the smart pirate do?

Answer
The smart pirate put two 17 pound and two 13 pound bars in his pack, which maximized his take at 60 pounds of gold compared to the greedy pirate's 50 pound load.

4.5.11. Suppose that $\left(F_{a_1} + F_{a_2} + \cdots + F_{a_{j-1}} + F_{a_j} + \cdots + F_{a_{r-1}} + F_{a_r}, b\right)$ is a winning Fibonacci Nim position. Show that if the "tail" of the Zeckendorf sum $k = F_{a_j} + \cdots + F_{a_{r-1}} + F_{a_r}$ satisfies $k \leq b$ and $2k < F_{a_{j-1}}$, then k coins can be safely taken from the pile of coins.

Answer
Removing k coins leaves the opponent with the position $\left(F_{a_1} + F_{a_2} + \cdots + F_{a_{j-1}}, 2k\right)$. This is a losing position since $2k < F_{a_{j-1}}$.

PROBLEM SET 4.6

4.6.1. Prove the identity $\sum_{k\geq 0} (-1)^k \binom{m}{k} \binom{n+k}{r} = (-1)^m \binom{n}{r-m}$.

[*Hint*: Start with two bags, the first with red balls numbered 1 through m and a second with green balls numbered 1 through n.]

Answer

D. Following the hint, consider two bags, one with red balls numbered 1 through m and another bag with green balls numbered 1 through n. Then the unsigned sum $\sum_{k\geq 0} \binom{m}{k} \binom{n+k}{r}$ counts the number of ways to choose a subset of k of the red balls, transfer them to the other bag, and then choose a subset of r of the balls from the second bag.

I. Let A be the subset of $m - k$ red balls left in the first bag and let B be the subset of red balls left in the second bag after r balls are removed from the second bag. Transferring the lowest numbered red ball in $A \cup B$ from one bag to the other is a sign reversing involution.

E. The exceptions occur when $A = B = \varnothing$. That is, all $k = m$ red balls were first moved to the second bag and all then all m of these balls were then removed from the second bag. The other $r - m$ balls removed from the second bag are a subset of the n green balls, so there are $\binom{n}{r-m}$ exceptional cases, each with the sign $(-1)^m$.

The sum of the alternating series is therefore $(-1)^m \binom{n}{r-m}$.

4.6.3. Let $s_n = \sum_{k=1}^{n} (-1)^{k+1} k(n+1-k)$. Use the interpretation that $k(n+1-k)$ gives the number of ways to place three square tiles on a $1 \times (n+2)$ board, with the middle tile on cell $k+1$. Arrangements with the middle tile on an even cell are counted positively. Now show that

(a) $s_{2m} = 0$, using the DIE method.

(b) $s_{2m-1} = m$

Answer

(a) Each arrangement with tiles on the three squares $(q, k+1, r)$ can be paired with the arrangement obtained by reflection over the midline of the board, as shown here.

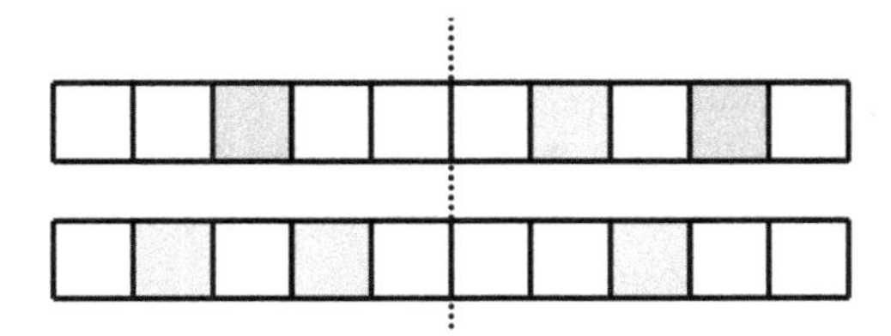

The middle tile of the reflected arrangement is on square $k' + 1 = 2m + 2 - k$ so k and k' have opposite parity. Reflection gives a sign-reversing involution of arrangements with no exceptional arrangements and therefore $s_{2m} = 0$.

(b) Let $n = 2m - 1$, so the board has $2m + 1$ squares. By part (a), the arrangements only using squares 1 through $2m$ contribute a 0 to the sum. Now consider arrangements of the form $(q, 2j - 1, 2m + 1)$ that have a tile on square $2m + 1$. Each of these negatively signed arrangements can be paired with the positively signed arrangement $(q, 2j, 2m + 1)$ for $1 \leq j \leq m$. The exceptional arrangements therefore have the form $(2j - 1, 2j, 2m + 1)$, for $j = 1, 2, \ldots, m$. Since there are m exceptional arrangements, all with positive sign, we conclude that $s_{2m-1} = m$.

4.6.5. The PIE formula can be written conveniently as

$$\left|A_1 \cup A_2 \cup \cdots \cup A_n\right| = S_1 - S_2 + \cdots + (-1)^{n+1} S_n$$

where

$$S_1 = \sum_{1 \leq i \leq n} |A_i|,\ S_2 = \sum_{1 \leq i < j \leq n} \left|A_i \cap A_j\right|,\ S_3 = \sum_{1 \leq i < j < k \leq n} \left|A_i \cap A_j \cap A_k\right|, \ldots,$$

$$S_n = \left|A_1 \cap A_2 \cap \cdots \cap A_n\right|.$$

Using this notation, show that

(a) $N_1 = S_1 - 2S_2 + 3S_3 - 4S_4 + \cdots + (-1)^{n-1}\, nS_n$ is the number of elements in $A_1 \cup A_2 \cup \cdots \cup A_n$ that belong to exactly one of the sets $A_1, \ldots, A_n$.

(b) $N_2 = S_2 - \binom{3}{2} S_3 + \binom{4}{2} S_4 - \binom{5}{2} S_5 + \cdots + (-1)^{n-2} \binom{n}{2} S_n$ is the number of elements in $A_1 \cup A_2 \cup \cdots \cup A_n$ that belong to exactly two of the sets $A_1, \ldots, A_n$.

Answer

(a) Suppose x is a member of just one set. Then x is counted once by S_1 and is not counted by any S_j for $j > 1$. Therefore, the element x is counted exactly once by the right side of the formula, as needed. Now suppose some element y is a member of k sets, where $k \geq 2$. Then y is counted $\binom{k}{j}$ times by S_j for $j = 1, 2, \ldots, k$. The contribution made by y to the right side of the formula is therefore

$$N_y = \binom{k}{1} - 2\binom{k}{2} + 3\binom{k}{3} - 4\binom{k}{4} + \cdots + (-1)^{k-1} k \binom{k}{k}.$$

But $0 = \sum_{j=1}^{k} (-1)^j j \binom{k}{j} = -N_y$ (See Example 4.2). Alternatively, differentiate $(1+x)^k = \sum_{j=0}^{k} \binom{k}{j} x^j$ and then set $x = -1$. Thus, any element y that belongs to more than one of the sets $A_1, \ldots, A_n$ contributes a 0 to the right side of the formula.

(b) Any element belonging to just one of the sets $A_1, \ldots, A_n$ is not counted by the right side of the given formula since only the terms S_j for $j > 1$ appear. Now suppose x is a member of exactly two sets. Then x is counted once by S_2 and is not counted by any S_j for $j > 2$. Therefore, the element x is counted exactly once by the right side of the formula, as needed. Now suppose some element y is a member of k sets, where $k \geq 3$. Then y is counted $\binom{k}{j}$ times by S_j for $j = 2, 3, \ldots, k$. The contribution made by y to the right side of the formula is therefore

$$N_y = \binom{k}{2} - \binom{3}{2}\binom{k}{3} + \binom{4}{2}\binom{k}{4} - \binom{5}{2}\binom{k}{5}$$

$$+ \cdots + (-1)^{k-2}\binom{k}{2}\binom{k}{k}.$$

But $N_y = \sum_{j=2}^{k} (-1)^j \binom{j}{2}\binom{k}{j} = (-1)^2 \delta_{2,k} = 0$ since $k \neq 2$ (see Theorem 4.5). Thus, any element y that belongs to more than two of the sets $A_1, \ldots, A_n$ contributes a 0 to the right side of the formula.

4.6.7. The number of elements that belong to exactly one of the four sets A_1, A_2, A_3, A_4 is given by $N_1 = S_1 - 2S_2 + 3S_3 - 4S_4$ (see Problem 4.6.5(a)), where

$$S_1 = \sum_{1\le i\le 4} |A_i|, S_2 = \sum_{1\le i<j\le 4} |A_i \cap A_j|, S_3 = \sum_{1\le i<j<k\le 4} |A_i \cap A_j \cap A_k|,$$

$$S_4 = |A_1 \cap A_2 \cap A_3 \cap A_4|$$

(a) Use the formula for N_1 to count the number of permutations of 4 elements with exactly one fixed point.

(b) Verify that the answer obtained in part (a) is $4D_3$.

Answer

(a) Let A_i be the set of permutations of [4] that fix i, and permute the remaining elements without restriction. Therefore, $|A_i| = 3!$, $|A_i \cap A_j| = 2!$, $|A_i \cap A_j \cap A_k| = 1$, $|A_1 \cap A_2 \cap A_3 \cap A_4| = 1$ and so

$$N_1 = (4 \cdot 3!) - 2(6 \cdot 2!) + 3(4 \cdot 1!) - 4 \cdot 0! = 24 - 24 + 12 - 4 = 8$$

(b) $4D_3 = 4 \cdot 2 = 8$.

4.6.9. Determine $r_2(\mathcal{B})$ and $r_3(\mathcal{B})$ for each of these boards $\mathcal{B}$ of forbidden squares.

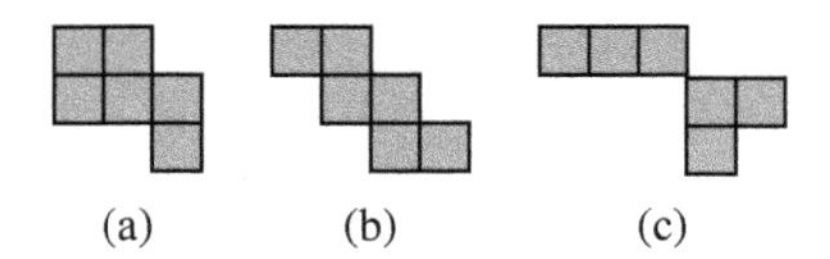

Answer

(a) 8, 2 **(b)** 10, 4 **(c)** 10, 3

4.6.11. Show that two identical nonattacking rooks can be placed on the "staircase" board of height n shown in $2n^2 - 3n + 1$ ways.

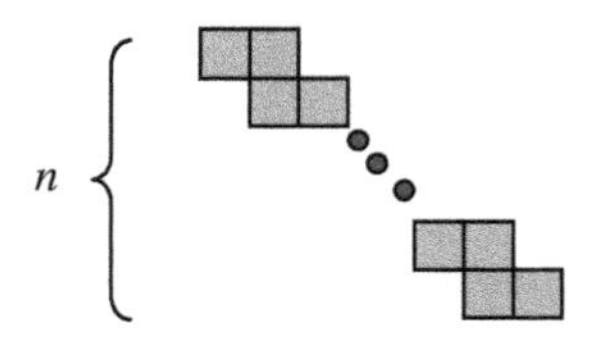

Answer

There are $\binom{2n}{2} = \frac{(2n)(2n-1)}{2} = 2n^2 - n$ ways to place two rooks on the board that include the n disallowed cases in which both rooks are in the same row and the $n - 1$ disallowed cases in which both rooks are in the same column. Therefore, there are $2n^2 - n - n - (n-1) = 2n^2 - 3n + 1$ nonattacking placements of the two rooks.

5

RECURRENCE RELATIONS

PROBLEM SET 5.2

5.2.1. Why should you not be surprised in Example 5.2 to discover that the number of binary sequences of length n with no two 1s adjacent is given by the combinatorial Fibonacci number f_{n+1}? Recall that f_{n+1} counts the number of tilings of a $1 \times (n+1)$ board with squares and dominoes.

Answer
First, extend any binary sequence of length n to a sequence of length $n+1$ that ends with a 0. Then replace each 1 0 by a domino and finally replace each remaining 0 with a square. This associates each binary sequence with no adjacent 1s with a tiling of a board of length $n+1$ with dominos and squares. Since the correspondence is reversible, this is a bijection of the binary sequences without consecutive 1s and the f_{n+1} tilings, and proves the counting formula.

5.2.3. In the block walking diagram below, there are two choices at each intersection: move to the next horizontal row below by turning to the southeast, or move directly south to the intersection two rows below.

Solutions Manual to Accompany Combinatorial Reasoning: An Introduction to the Art of Counting, First Edition. Duane DeTemple and William Webb.

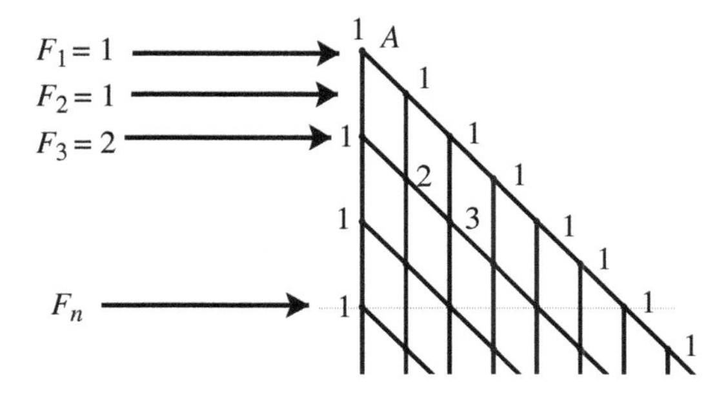

Let F_n denote the total number of paths that reach row n that start at point A in row 1. [*Hint*: F_n also gives the number of paths that move from a starting point in row r to a point in row $r + n - 1$.]

(a) Is the notation used in the diagram appropriate?

(b) Give a block walking proof of the identity $F_{m+n} = F_m F_{n+1} + F_{m-1} F_n$.

(c) Use the identity from part (b) to prove that F_m divides F_{km} for all $k \geq 0$ and $m \geq 1$.

Answer

(a) Any of the F_n paths that reach row n either start with a diagonal southeast move from A or with a vertically downward move that skips over a row. There are then, respectively, F_{n-1} and F_{n-2} ways to complete the paths to row n. Therefore, $F_n = F_{n-1} + F_{n-2}$, the Fibonacci recurrence. Since $F_1 = F_2 = 1$, we conclude that F_n is the nth Fibonacci number as the notation suggests.

(b) There are two types of walks that reach row $n + m$, distinguished by whether or not a point along row m is a point of the path or else row m is skipped over. There are F_m ways to reach a point of row m, and from each of these points there are F_{n+1} ways to reach row $n + m$. Thus, $F_m\ F_{n+1}$ paths that reach row $n + m$ pass through a point of row m. Similarly, there are F_{m-1} ways that first reach a point of row $m - 1$, then skip over row m by making a downward move, and then from each of these points continue on to row $m + n$ in F_n ways. Thus, $F_{m-1}\ F_n$ paths to row $n + m$ do not pass through a point of row m. Altogether, we see that $F_{m+n} = F_m F_{n+1} + F_{m-1} F_n$.

(c) For $n = km$, the identity becomes $F_{m+km} = F_m F_{km+1} + F_{m-1} F_{km}$. Now use mathematical induction. Certainly, F_m divides F_m, so the result holds for $k = 1$. Now suppose that F_m divides F_{km}. Then F_m divides $F_m F_{km+1} + F_{m-1} F_{km}$ and so it divides F_{m+km} by the identity.

5.2.5. In Fibonacci's influential book *Liber Abaci,* he proposed this problem:

"A certain man put a pair of rabbits in a place surrounded by a wall. If the rabbits can breed during January, how many pairs of rabbits can be produced in a year, if it is supposed that every month each pair produces a new pair which can breed from the second month?" Find a recurrence relation and then solve Fibonacci's rabbit problem.

Answer

Let h_n denote the number of pairs after n months, so $h_1 = 1$ and $h_2 = 1$. After 3 months, there is the original pair and a new pair so $h_3 = 2$. For $n \geq 1$, the h_{n+2} pairs are either the h_{n+1} pairs in existence after $n + 1$ months, or the h_n new pairs born to pairs in existence after n months. Thus $h_{n+2} = h_{n+1} + h_n$.

5.2.7. Prove the identity $F_0 + F_2 + F_4 + \cdots + F_{2m} = F_{2m+1} - 1, m \geq 0$, by

(a) mathematical induction.

(b) summing $F_{2n+1} - F_{2n-1} = F_{2n}$.

Answer

(a) The identity holds for $m = 0$ since $F_0 = 0 = 1 - 1 = F_1 - 1$. Now assume the formula hold some $m \geq 0$. Then $F_0 + F_2 + F_4 + \cdots + F_{2m} + F_{2m+2} = F_{2m+1} + F_{2m+2} - 1 = F_{2m+3} - 1$ so the identity holds for all $m \geq 0$ by the principle of mathematical induction.

(b) $F_0 + F_2 + F_4 + \cdots + F_{2m} = 0 + (F_3 - F_1) + (F_5 - F_3) + (F_7 - F_5) + \cdots + (F_{2m+1} - F_{2m-1}) = -F_1 + F_{2m+1} = F_{2m+1} - 1.$

5.2.9. Determine a simple expression involving one Lucas number for each of these sums of Lucas numbers.

(a) $L_0 + L_1 + L_2 + \cdots + L_m$

(b) $L_0 + L_2 + L_4 + \cdots + L_{2m}$

(c) $L_1 + L_3 + L_5 + \cdots + L_{2m-1}$

Answer

Using the telescoping sum method

(a) $L_0 + L_1 + L_2 + L_3 + \cdots + L_m = (L_2 - L_1) + (L_3 - L_2) + (L_4 - L_3) + (L_5 - L_4) + \cdots + (L_{m+2} - L_{m+1}) = -L_1 + L_{m+2} = -1 + L_{m+2}$

(b) $L_0 + L_2 + L_4 + L_6 + \cdots + L_{2m} = L_0 + (L_3 - L_1) + (L_5 - L_3) + (L_7 - L_5) + \cdots + (L_{2m+1} - L_{2m-1}) = L_0 - L_1 + L_{2m+1} = 2 - 1 + L_{2m+1} = L_{2m+1} + 1$

(c) $L_1 + L_3 + L_5 + L_7 + \cdots + L_{2m-1} = (L_2 - L_0) + (L_4 - L_2) + (L_6 - L_4) + \cdots + (L_{2m} - L_{2m-2}) = -L_0 + L_{2m} = -2 + L_{2m} = L_{2m} - 2$

The identities also follow by mathematical induction.

5.2.11. If i is the annual percentage yield, then the principal is doubled in $N = \lceil f(i) \rceil$ years where $2 = \left(1 + \frac{i}{100}\right)^{f(i)}$ (see Example 5.7). Graph both $y = f(x)$ and $y = 72/x$ to see how accurately the "rule of 72" estimates doubling times of an investment growing at the annual interest rate $x = i$.

[*Hint*: Use the logarithm to solve for the function $f(i)$.]

Answer

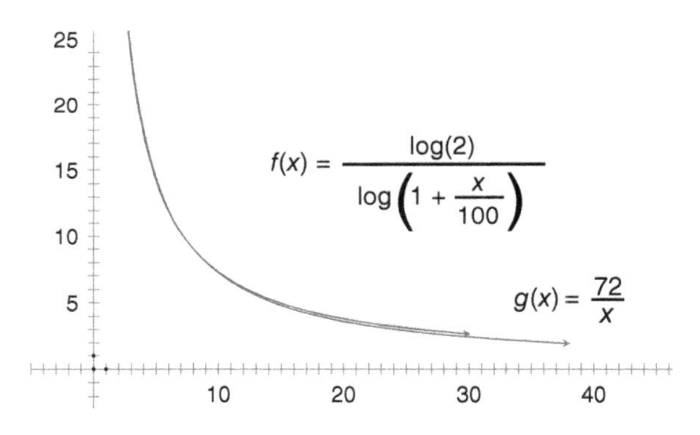

5.2.13. Revise Problem 5.2.12 to define the Lucas numbers L_{-k} and prove a formula that expresses how L_{-k} is related to L_k for all $k \geq 0$.

Answer

Setting $n = 0$ and 1 in the recurrence $L_{n-1} = L_{n+1} - L_n$ shows that $L_{-1} = L_1 - L_0 = 1 - 2 = -1 = -L_1$ and $L_{-2} = L_0 - L_{-1} = 2 - (-1) = 3 = L_2$. It seems that $L_{-k} = (-1)^k L_k$. Using mathematical induction, the formula is true for $k = 1$ and $k = 2$, so assume that $L_{-k} = (-1)^k L_k$ and $L_{-(k+1)} = (-1)^{k+1} L_{k+1}$ Then $L_{-(k+2)} = L_{-k} - L_{-(k+1)} = (-1)^k L_k - (-1)^{k+1} L_{k+1} = (-1)^{k+2} (L_k + L_{k+1}) = (-1)^{k+2} L_{k+2}$.

5.2.15. Prove these identities.

(a) $\varphi^n = \varphi F_n + F_{n-1}$

(b) $\hat{\varphi}^n = \hat{\varphi} F_n + F_{n-1}$

[*Hint*: Apply the operator $C(E) = E^2 - E - 1$ to the sequences $h_n = \varphi^n - \varphi F_n - F_{n-1}$ and $g_n = \hat{\varphi}^n - \hat{\varphi} F_n - F_{n-1}$.]

Answer

(a) Let $h_n = \varphi^n - \varphi F_n - F_{n-1}$. Then $h_1 = \varphi^1 - \varphi F_1 - F_0 = \varphi - \varphi = 0$ and $h_2 = \varphi^2 - \varphi F_2 - F_1 = \varphi^2 - \varphi - 1 = 0$. But by the linearity of the operator $C(E)$ we also have $C(E) h_n = \left(E^2 - E - 1\right) \varphi^n - \varphi \left(E^2 - E - 1\right) F_n - \left(E^2 - E - 1\right) F_{n-1} = 0 - 0 - 0 = 0$, so $h_n = 0$ for all $n \geq 1$.

(b) Let $g_n = \hat{\varphi}^n - \hat{\varphi} F_n - F_{n-1}$. Then $g_1 = \hat{\varphi}^1 - \hat{\varphi} F_1 - F_0 = \hat{\varphi} - \hat{\varphi} = 0$ and $g_2 = \hat{\varphi}^2 - \hat{\varphi} F_2 - F_1 = \hat{\varphi}^2 - \hat{\varphi} - 1 = 0$. Also, $C(E) h_n = \left(E^2 - E - 1\right) \hat{\varphi}^n - \hat{\varphi} \left(E^2 - E - 1\right) F_n - \left(E^2 - E - 1\right) F_{n-1} = 0 - 0 - 0 = 0$, so $g_n = 0$ for all $n \geq 1$.

5.2.17. Use the Binet formulas to prove these identities.

(a) $F_{n+1} = \dfrac{F_n + L_n}{2}$

(b) $L_{n+1} = \dfrac{5F_n + L_n}{2}$

Answer

(a)
$$F_n + L_n = \frac{\varphi^n - \hat{\varphi}^n}{\sqrt{5}} + \left(\varphi^n + \hat{\varphi}^n\right) = \varphi^n \left(\frac{1+\sqrt{5}}{\sqrt{5}}\right) - \hat{\varphi}^n \left(\frac{1-\sqrt{5}}{\sqrt{5}}\right)$$
$$= \varphi^n \left(\frac{2\varphi}{\sqrt{5}}\right) - \hat{\varphi}^n \left(\frac{2\hat{\varphi}}{\sqrt{5}}\right) = 2\frac{\varphi^{n+1} - \hat{\varphi}^{n+1}}{\sqrt{5}} = 2F_{n+1}$$

(b)
$$5F_n + L_n = 5\frac{\varphi^n - \hat{\varphi}^n}{\sqrt{5}} + \left(\varphi^n + \hat{\varphi}^n\right) = \left(1 + \sqrt{5}\right) \varphi^n + \left(1 - \sqrt{5}\right) \hat{\varphi}^n$$
$$= 2\varphi^{n+1} + 2\hat{\varphi}^{n+1} = 2L_{n+1}$$

5.2.19. Use the Binet formula to prove these identities.

(a) $F_n^2 + F_{n+1}^2 = F_{2n+1}$

(b) $L_n^2 + L_{n+1}^2 = 5F_{2n+1}$

(c) $F_{n+1}^2 - F_n^2 = \dfrac{L_{2n+1} + 4(-1)^n}{5}$

(d) $L_n^2 - 5F_n^2 = 4(-1)^n$

Answer

Note that $\varphi^2 = \varphi + 1, \hat{\varphi}^2 = \hat{\varphi} + 1, \varphi\hat{\varphi} = -1, \varphi + \hat{\varphi} = 1, \varphi - \hat{\varphi} = \sqrt{5}$.

(a)
$$\begin{aligned}
F_n^2 + F_{n+1}^2 &= \left(\frac{\varphi^n - \hat{\varphi}^n}{\sqrt{5}}\right)^2 + \left(\frac{\varphi^{n+1} - \hat{\varphi}^{n+1}}{\sqrt{5}}\right)^2 \\
&= \frac{\varphi^{2n} - 2(-1)^n + \hat{\varphi}^{2n} + \varphi^{2n+2} - 2(-1)^{n+1} + \hat{\varphi}^{2n+2}}{5} \\
&= \frac{\varphi^{2n+2} + \varphi^{2n} + \hat{\varphi}^{2n+2} + \hat{\varphi}^{2n}}{5} \\
&= \frac{\varphi^{2n+1}(\varphi - \hat{\varphi}) + \hat{\varphi}^{2n+1}(\hat{\varphi} - \varphi)}{5} \\
&= \sqrt{5}\frac{\varphi^{2n+1} - \hat{\varphi}^{2n+1}}{5} = F_{2n+1}
\end{aligned}$$

(b)
$$\begin{aligned}
L_n^2 + L_{n+1}^2 &= (\varphi^n + \hat{\varphi}^n)^2 + \left(\varphi^{n+1} + \hat{\varphi}^{n+1}\right)^2 \\
&= \varphi^{2n} + 2(-1)^n + \hat{\varphi}^{2n} + \varphi^{2n+2} + 2(-1)^{n+1} + \hat{\varphi}^{2n+2} \\
&= \varphi^{2n+2} + \varphi^{2n} + \hat{\varphi}^{2n+2} + \hat{\varphi}^{2n} \\
&= \varphi^{2n+1}(\varphi - \hat{\varphi}) + \hat{\varphi}^{2n+1}(\hat{\varphi} - \varphi) \\
&= \sqrt{5}\left(\varphi^{2n+1} - \hat{\varphi}^{2n+1}\right) = 5F_{2n+1}
\end{aligned}$$

(c)
$$\begin{aligned}
F_{n+1}^2 - F_n^2 &= \left(\frac{\varphi^{n+1} - \hat{\varphi}^{n+1}}{\sqrt{5}}\right)^2 - \left(\frac{\varphi^n - \hat{\varphi}^n}{\sqrt{5}}\right)^2 \\
&= \frac{\varphi^{2n+2} - 2(-1)^{n+1} + \hat{\varphi}^{2n+2} - \varphi^{2n} + 2(-1)^n - \hat{\varphi}^{2n}}{5} \\
&= \frac{\varphi^{2n+2} - \varphi^{2n} + \hat{\varphi}^{2n+2} - \hat{\varphi}^{2n} + 4(-1)^n}{5} \\
&= \frac{\varphi^{2n+1} + \hat{\varphi}^{2n+1} + 4(-1)^n}{5} = \frac{L_{2n+1} + 4(-1)^n}{5}
\end{aligned}$$

(d)
$$\begin{aligned}
L_n^2 - 5F_n^2 &= (\varphi^n + \hat{\varphi}^n)^2 - 5\left(\frac{\varphi^n - \hat{\varphi}^n}{\sqrt{5}}\right)^2 \\
&= (\varphi^{2n} + 2(-1)^n + \hat{\varphi}^{2n}) - (\varphi^{2n} - 2(-1)^n + \hat{\varphi}^{2n}) = 4(-1)^n
\end{aligned}$$

5.2.21. Prove that

(a) $\begin{bmatrix} F_{n+1} & F_n \\ F_n & F_{n-1} \end{bmatrix} = \begin{bmatrix} 1 & 1 \\ 1 & 0 \end{bmatrix}^n$

(b) $F_{n+1}F_{n-1} - F_n^2 = (-1)^n$

Answer

(a) Use mathematical induction. For $n = 1$, the formula is true since $F_0 = 0$ and $F_1 = F_2 = 1$. Now assume the formula is true for some $n \geq 1$. Then $\begin{bmatrix} 1 & 1 \\ 1 & 0 \end{bmatrix}^{n+1} = \begin{bmatrix} 1 & 1 \\ 1 & 0 \end{bmatrix}\begin{bmatrix} 1 & 1 \\ 1 & 0 \end{bmatrix}^n = \begin{bmatrix} 1 & 1 \\ 1 & 0 \end{bmatrix}\begin{bmatrix} F_{n+1} & F_n \\ F_n & F_{n-1} \end{bmatrix} = \begin{bmatrix} F_{n+1} + F_n & F_n + F_{n-1} \\ F_{n+1} & F_n \end{bmatrix} = \begin{bmatrix} F_{n+2} & F_{n+1} \\ F_{n+1} & F_n \end{bmatrix}$.

(b) $F_{n+1}F_{n-1} - F_n^2 = \det\begin{bmatrix} F_{n+1} & F_n \\ F_n & F_{n-1} \end{bmatrix} = \det\begin{bmatrix} 1 & 1 \\ 1 & 0 \end{bmatrix}^n = \left(\det\begin{bmatrix} 1 & 1 \\ 1 & 0 \end{bmatrix}\right)^n = (-1)^n$.

PROBLEM SET 5.3

5.3.1. Let $h_n = 2h_{n-1} + 3h_{n-2},\ n \geq 2$.

(a) What are the corresponding eigenvalues?

(b) What is the general solution (power sum) with unspecified initial conditions?

(c) Solve the recurrence relation with the initial conditions $h_0 = 0$ and $h_1 = 1$.

(d) Solve the recurrence relation with the initial conditions $h_0 = 3$ and $h_1 = 1$.

Answer

(a) $\alpha = 3$ and $\beta = -1$

(b) $c_1 3^n + c_2 (-1)^n$

(c) $\dfrac{3^n - (-1)^n}{4}$

(d) $3^n + 2(-1)^n$

5.3.3. Let $h_n = h_{n-1} - 3h_{n-2},\ n \geq 2$.

(a) What are the corresponding eigenvalues?

(b) What is the general solution (power sum) with unspecified initial conditions?

Answer

(a) $\alpha = \dfrac{1 + i\sqrt{11}}{2}$ and $\beta = \dfrac{1 - i\sqrt{11}}{2}$

(b) $c_1 \left(\dfrac{1 + i\sqrt{11}}{2} \right)^n + c_2 \left(\dfrac{1 - i\sqrt{11}}{2} \right)^n$

5.3.5. Let $h_n = 2h_{n-1} - h_{n-2}$, $n \geq 2$.

(a) What are the corresponding eigenvalues?

(b) What is the general solution (power sum) with unspecified initial conditions?

Answer

(a) $\alpha = \beta = 1$

(b) $c_1 + c_2 n$

5.3.7. Verify that every linear polynomial $c_0 + c_1 n$ is annihilated by $(E - 1)^2$.

Answer

$$\begin{aligned}(E-1)^2 \left(c_0 + c_1 n\right) &= (E-1)\left(c_0 + c_1 (n+1) - c_0 - c_1 n\right) \\ &= (E-1)\left(c_1\right) = c_1 - c_1 = 0\end{aligned}$$

5.3.9. **(a)** What recurrence relation has the eigenvalues φ^2 and $\hat{\varphi}^2$?

(b) Why can the general solution of the recurrence relation be written as $c_1 F_{2n} + c_2 L_{2n}$?

Answer

(a) $\left(E - \varphi^2\right)\left(E - \hat{\varphi}^2\right) = E^2 - \left(\varphi^2 + \hat{\varphi}^2\right) E + (\varphi\hat{\varphi})^2 = E^2 - L_2 E + 1 = E^2 - 3E + 1$, so the recurrence relation is $h_{n+2} = 3h_{n+1} - h_n$.

(b) By the Binet formulas for the Fibonacci and Lucas numbers, $F_{2n} = \dfrac{\varphi^{2n} - \hat{\varphi}^{2n}}{\sqrt{5}}$ and $L_{2n} = \varphi^{2n} + \hat{\varphi}^{2n}$, so the recurrence relation $h_{n+2} = 3h_{n+1} - h_n$ is solved by $h_n = c_1 F_{2n} + c_2 L_{2n}$ This is the general solution since there are unique coefficients c_1 and c_2 determined by any initial conditions $h_0 = 2c_2$ and $h_1 = c_1 + 3c_2$. Indeed, $c_2 = \dfrac{h_0}{2}$ and $c_1 = h_1 - \dfrac{3}{2} h_0$.

5.3.11. Let α and β be the eigenvalues of the characteristic polynomial $C(x) = x^2 - a_1 x - a_2$. Show that $\alpha + \beta = a_1$ and $\alpha\beta = -a_2$.

Answer

$C(x) = x^2 - a_1x - a_2 = (x - \alpha)(x - \beta) = x^2 - (\alpha + \beta)x + \alpha\beta$, so equating like coefficients of powers of x gives $\alpha + \beta = a_1$ and $\alpha\beta = -a_2$.

5.3.13. **(a)** Show that if $g_n = c_1\alpha^n + c_2\beta^n$, where α and β are distinct eigenvalues, then the constants c_1 and c_2 are uniquely determined by the initial conditions $g_0 = A$ and $g_1 = B$.

(b) Suppose $g_n = c_1\alpha^n + c_2n\alpha^n$ is the general solution of a second order recurrence relation for which $\alpha \neq 0$ is a repeated eigenvalue. Show that the constants c_0 and c_1 are uniquely determined by the initial conditions $g_0 = A$ and $g_1 = B$.

Answer

(a) In matrix form, we must solve $\begin{bmatrix} 1 & 1 \\ \alpha & \beta \end{bmatrix} \begin{bmatrix} c_1 \\ c_2 \end{bmatrix} = \begin{bmatrix} A \\ B \end{bmatrix}$. This equation is uniquely solvable since $\det \begin{bmatrix} 1 & 1 \\ \alpha & \beta \end{bmatrix} = \beta - \alpha \neq 0$; that is, the matrix is nonsingular (invertible).

(b) In matrix form, we must solve $\begin{bmatrix} 1 & 0 \\ \alpha & \alpha \end{bmatrix} \begin{bmatrix} c_0 \\ c_1 \end{bmatrix} = \begin{bmatrix} A \\ B \end{bmatrix}$. Since $\det \begin{bmatrix} 1 & 0 \\ \alpha & \alpha \end{bmatrix} = \alpha \neq 0$, the matrix is invertible and there is a unique solution of the matrix equation.

5.3.15. Show that the Pell number P_{n+1} gives the number of ways to tile a $1 \times n$ board with squares and dominoes, where the squares are either red or blue and the dominoes are all white.

Answer

Let h_n denote the number of ways to tile a board of length n. There are $h_1 = 2$ ways to tiles a board of length 1 since the square tile can be red or blue. There are $h_2 = 5$ ways to tile a board of length 2, since there are $2^2 = 4$ ways that use squares and one additional way with a domino. A tiling of a board of length n can end with a square of either of two colors, preceded by a tiling of a board of length $n - 1$. There are also the tilings that end with a domino and are preceded by a tiling of a board of length $n - 2$. Therefore, we have the homogeneous recurrence relation $h_n = 2h_{n-1} + h_{n-2}, n \geq 3$. This is the recurrence for the Pell numbers seen in Example 5.14, but with the shifted initial conditions $h_1 = 2$ and $h_2 = 5$.

Thus $h_n = P_{n+1} = \dfrac{\left(1 + \sqrt{2}\right)^{n+1} - \left(1 - \sqrt{2}\right)^{n+1}}{2\sqrt{2}}$.

5.3.17. Prove that

(a) the Pell numbers P_n are given by

$$\begin{bmatrix} P_{n+1} & P_n \\ P_n & P_{n-1} \end{bmatrix} = \begin{bmatrix} 2 & 1 \\ 1 & 0 \end{bmatrix}^n, n \geq 1.$$

(b) $P_{n+1}P_{n-1} - P_n^2 = (-1)^n$.

Answer

(a) Use mathematical induction: The result is true for $n = 1$ since $\begin{bmatrix} P_2 & P_1 \\ P_1 & P_0 \end{bmatrix} = \begin{bmatrix} 2 & 1 \\ 1 & 0 \end{bmatrix}$. Now assume that $\begin{bmatrix} P_{n+1} & P_n \\ P_n & P_{n-1} \end{bmatrix} = \begin{bmatrix} 2 & 1 \\ 1 & 0 \end{bmatrix}^n$ holds for an $n \geq 1$. Then

$$\begin{bmatrix} 2 & 1 \\ 1 & 0 \end{bmatrix}^{n+1} = \begin{bmatrix} 2 & 1 \\ 1 & 0 \end{bmatrix}\begin{bmatrix} 2 & 1 \\ 1 & 0 \end{bmatrix}^n = \begin{bmatrix} 2 & 1 \\ 1 & 0 \end{bmatrix}\begin{bmatrix} P_{n+1} & P_n \\ P_n & P_{n-1} \end{bmatrix}$$
$$= \begin{bmatrix} 2P_{n+1} + P_n & 2P_n + P_{n-1} \\ P_{n+1} & P_n \end{bmatrix} = \begin{bmatrix} P_{n+2} & P_{n+1} \\ P_{n+1} & P_n \end{bmatrix}$$

so the formula holds for $n+1$. By mathematical induction, the formula holds for all $n \geq 1$.

(b) The formula follows by taking determinants in the equation in part (a).

5.3.19. **(a)** Use a block walking argument (see Example 5.14) to show that $P_{m+n} = P_m P_{n+1} + P_{m-1}P_n$. [*Hint*: The paths from point A to row $m + n$ are of two types: those that pass through a point of row m and those that move directly downward from row $m - 1$ to row $m + 1$.]

(b) Use part (a) to prove that $P_{2n+2} = P_{n+1}P_{n+2} + P_nP_{n+1}$.

(c) Use the result of Problem 5.3.18(b) to prove that P_{n+1} divides $\sum_{k=0}^{n} P_{2k+1}$.

[*Hint*: What is the parity of $P_{n+2} + P_n$?]

Answer

(a) There are P_{m+n} paths from point A in row 1 to the points in row $m + n$. These paths can be partitioned into two types: Type 1, those passing through a point in row m, and Type 2, those that skip row m by moving directly downward from row $m-1$ to row $m+1$. There are P_m paths from A to row m, and from each of these points there are another P_{n+1} paths that complete the path to row $m+n$. This means there are $P_m P_{n+1}$ paths of Type 1. There are also P_{m-1} paths that reach row $m-1$,then move directly downward, and then complete the path to row $m+n$ in P_n ways This means there are $P_{m-1}P_n$ paths of Type 2. Summing the paths of the two types gives the identity.

(b) The identity follows if both m and n are replaced with $n+1$ to get $P_{2n+2} = P_{n+1}P_{n+2} + P_n P_{n+1} = P_{n+1}\left(P_{n+2} + P_n\right)$.

(c) From part (b) it is immediately clear that P_{n+1} divides P_{2n+2}. But since the Pell numbers alternate in parity, $P_{n+2} + P_n$ is even, say $P_{n+2} + P_n = 2R_n$. Then from part (b) we see that $2R_n\, P_{n+1} = P_{2n+2}$. But from Problem 5.3.18(b), $P_{2n+2} = 2\sum_{k=0}^{n} P_{2k+1}$, so $R_n P_{n+1} = \sum_{k=0}^{n} P_{2k+1}$ which shows that P_{n+1} divides $\sum_{k=0}^{n} P_{2k+1}$.

5.3.21. **(a)** Prove that the limit of the ratio of successive Pell numbers is the silver ratio. That is, prove that

$$\lim_{n\to\infty} \frac{P_{n+1}}{P_n} = 1 + \sqrt{2}.$$

(b) Prove that $\lim_{n\to\infty} \dfrac{P_n + P_{n-1}}{P_n} = \sqrt{2}$

Answer

(a) The Binet formula for the Pell number is $P_n = \dfrac{\alpha^n - \beta^n}{2\sqrt{2}}$, where $\alpha = 1 + \sqrt{2} \approx 2.4$ and $\beta = 1 - \sqrt{2} \approx -0.4$. Therefore, $\left|\dfrac{\beta}{\alpha}\right| < 1$ so $\lim_{n\to\infty}\left|\dfrac{\beta}{\alpha}\right|^n = 0$ and it follows that $\lim_{n\to\infty}\dfrac{P_{n+1}}{P_n} = \lim_{n\to\infty}\dfrac{\alpha^{n+1}}{\alpha^n}\dfrac{1-(\beta/\alpha)^{n+1}}{1-(\beta/\alpha)^n} = \alpha\dfrac{1-0}{1-0} = \alpha = 1+\sqrt{2}$.

(b) $\lim_{n\to\infty}\dfrac{P_n + P_{n-1}}{P_n} = \lim_{n\to\infty}\dfrac{P_{n+1} - P_n}{P_n} = \lim_{n\to\infty}\left(\dfrac{P_{n+1}}{P_n} - 1\right) = \left(1+\sqrt{2}\right) - 1 = \sqrt{2}$.

5.3.23. Prove the following recurrences for the Pell and half companion Pell numbers defined in Problem 5.3.22.

(a) $H_{n+1} = H_n + 2P_n$ for all $n \geq 0$

(b) $P_{n+1} = H_n + P_n$ for all $n \geq 0$

Answer

(a) The sequence $g_n = H_{n+1} - H_n - 2P_n$ is annihilated by the Pell recurrence operator $E^2 - 2E - 1$ and has the initial conditions $g_0 = H_1 - H_0 - 2P_0 = 1 - 1 - 0 = 0$ and $g_1 = H_2 - H_1 - 2P_1 = 3 - 1 - 2 = 0$. Therefore g_n is the identically 0 sequence.

(b) The sequence $h_n = P_{n+1} - H_n - P_n$ is annihilated by the Pell recurrence operator $E^2 - 2E - 1$ and has the initial conditions $h_0 = P_1 - H_0 - P_0 = 1 - 1 - 0 = 0$ and $h_1 = P_2 - H_1 - P_1 = 2 - 1 - 1 = 0$. Therefore h_n is the identically 0 sequence.

PROBLEM SET 5.4

5.4.1. Let $h_n = 2h_{n-1} + 3h_{n-2} - 6h_{n-3},\ n \geq 3$.

(a) What are the corresponding eigenvalues?

(b) What is the general solution (GPS) with unspecified initial conditions?

Answer

(a) $\alpha = \sqrt{3}, \beta = -\sqrt{3}$, and $\gamma = 2$

(b) $c_1 \left(\sqrt{3}\right)^n + c_2 \left(-\sqrt{3}\right)^n + c_3 2^n$

5.4.3. Suppose that a sequence satisfies a linear homogeneous recurrence relation of order 5. Support your answers to these questions.

(a) Could it satisfy a recurrence of order 3?

(b) Must it satisfy a recurrence of order 7?

Answer

(a) It is possible. The fifth order annihilating operator $C(E)$ may have a third order divisor that annihilates the sequence. (See Example 5.19.)

(b) Yes. If $P(E)$ is any second order operator, then the seventh-order operator $P(E)\,C(E)$ continues to annihilate the sequence.

5.4.5. **(a)** Show that the recurrence relation $h_{n+3} = a_1 h_{n+2} + a_2 h_{n+1} + a_3 h_n$ can be written as the equivalent matrix equation

$$\begin{bmatrix} h_{n+3} \\ h_{n+2} \\ h_{n+1} \end{bmatrix} = M \begin{bmatrix} h_{n+2} \\ h_{n+1} \\ h_n \end{bmatrix}$$

where

$$M = \begin{bmatrix} a_1 & a_2 & a_3 \\ 1 & 0 & 0 \\ 0 & 1 & 0 \end{bmatrix}$$

is the *companion matrix* of the characteristic polynomial

$$C(x) = x^3 - a_1x^2 - a_2x - a_3.$$

(b) Show that $C(x)$ is the characteristic polynomial of M.

Answer

(a) $\begin{bmatrix} a_1 & a_2 & a_3 \\ 1 & 0 & 0 \\ 0 & 1 & 0 \end{bmatrix}\begin{bmatrix} h_{n+2} \\ h_{n+1} \\ h_n \end{bmatrix} = \begin{bmatrix} a_1h_{n+2} + a_2h_{n+1} + a_3h_n \\ h_{n+2} \\ h_{n+1} \end{bmatrix} = \begin{bmatrix} h_{n+3} \\ h_{n+2} \\ h_{n+1} \end{bmatrix}$

(b) $\det(xI - M) = \det\begin{bmatrix} x - a_1 & -a_2 & -a_3 \\ -1 & x & 0 \\ 0 & -1 & x \end{bmatrix}$

$$= \left(x - a_1\right)x^2 - a_2x - a_3 = x^3 - a_1x^2 - a_2x - a_3$$

5.4.7. What GPS (generalized power sum) is annihilated by $(E-2)^2(E+1)^3$?

Answer

$$c_0 2^n + c_1 n 2^{n-1} + d_0(-1)^n + d_1\binom{n}{1}(-1)^{n-1} + d_2\binom{n}{2}(-1)^{n-2}$$

5.4.9. Solve the recurrence relation $h_{n+4} = 6h_{n+2} - 8h_{n+1} + 3h_n$, where $h_0 = 1, h_1 = 0, h_2 = 21, h_3 = 0$.

Answer

The characteristic polynomial $C(x) = x^4 - 6x^2 + 8x - 3 = (x-1)^3(x+3)$ shows that the general solution is $h_n = c_0 + c_1 n + c_2\binom{n}{2} + d_0(-3)^n$. The initial conditions are

$$h_0 = 1 = c_0 + d_0, h_1 = 0 = c_0 + c_1 - 3d_0,$$
$$h_2 = 21 = c_0 + 2c_1 + c_2 + 9d_0, h_3 = 0 = c_0 + 3c_1 + 3c_2 - 27d_0$$

Solving for the constants gives $c_0 = 0, c_1 = 3, c_2 = 6, d_0 = 1$ so $h_n = 3n + 6\binom{n}{2} + (-3)^n = 3n^2 + (-3)^n$.

5.4.11. **(a)** Compute the next 12 Perrin numbers of Example 5.21 in the table started below.

n	0	1	**2**	**3**	4	**5**	6	**7**	8	9	10	**11**
P_n	3	0	**2**	**3**	2	**5**	5	**7**	10	12	17	**22**

(b) What property seems to be true when n is a prime number?

(c) Does the converse hold? Investigate the Perrin numbers with an Internet search.

Answer

(a)

12	**13**	14	15	16	**17**	18	**19**	20	21	22	**23**
29	**39**	51	68	90	**119**	158	**209**	277	367	486	**644**

(b) When n is prime it divides p_n.

(c) The converse of the property does not hold, since it has been shown there are Perrin *pseudoprimes;* that is, there exist composite Perrin numbers p_n that are divisible by n. The smallest is when $n = 521^2$, discovered by Adams and Shank in 1982. In 2010, it was shown by Jon Grantham there are infinitely many Perrin pseudoprimes.

5.4.13. The nth *Padovan number* v_n is the number of compositions (ordered sums) of n for which the summands are odd integers ≥ 3. For example, $v_{10} = 3$ since 10 has the 3 compositions 3 + 7, 7 + 3, and 5 + 5: Notice that summands can be repeated and their order matters. Show that the Padovan numbers satisfy the Perrin recursion $v_{n+3} = v_{n+1} + v_n$, $n \geq 0$ investigated in Example 5.21.

(Note: It can be shown that

$$v_n = \frac{(\beta-1)(\gamma-1)}{(\beta-\alpha)(\gamma-\alpha)}\alpha^n + \frac{(\alpha-1)(\gamma-1)}{(\alpha-\beta)(\gamma-\beta)}\beta^n + \frac{(\alpha-1)(\beta-1)}{(\alpha-\gamma)(\beta-\gamma)}\gamma^n$$

where α, β, γ are the eigenvalues of the Perrin recursion.)

Answer

Each such composition either begins with a 3 or with a larger number. Each composition of n+3 beginning with a 3 is obtained from a composition of n by adding the first term of 3. Each composition not beginning with a 3 is obtained from a composition of n +1 by adding 2 to the first term. This correspondence is one to one.

5.4.15. Suppose that the sequences u_n and v_n satisfy the coupled linear recurrences $P(E)\,u_n + Q(E)\,v_n = 0$ and $R(E)\,u_n + S(E)\,v_n = 0$, where $P, Q, R,$ and S are polynomials. Prove that if $T(x) = P(x)\,S(x) - Q(x)\,R(x)$ then $T(E)\,u_n = T(E)\,v_n = 0$, so u_n and v_n satisfy a common linear recurrence of order k, where k is the degree of the polynomial T.

Answer

$$\begin{aligned} T(E)\,u_n &= (P(E)\,S(E) - Q(E)\,R(E))\,u_n \\ &= S(E)\,P(E)\,u_n - Q(E)\,R(E)\,u_n \\ &= -S(E)\,Q(E)\,v_n + Q(E)\,S(E)\,v_n = 0. \end{aligned}$$

An analogous calculation shows that $T(E)\,v_n = 0$.

5.4.17. Let u_n denote the number of ways to tile a $2 \times n$ board with squares and horizontal dominoes, and let v_n denote the number of ways to tile a $2 \times n$ pruned board that has a missing corner cell.

(a) Show that $u_{n+2} = u_{n+1} + u_n + 2v_{n+1}$ and $v_{n+1} = u_n + v_n$.

(b) Show that both u_n and v_n satisfy the order three recurrence $h_{n+3} = 2h_{n+2} + 2h_{n+1} - h_n$.

[*Hint*: Use the result of Problem 5.4.15.]

(c) Explain why F_{n+1}^2 and F_nF_{n+1}satisfy the recurrence of part (b).

Answer

(a) Any tiling of a board of length $n + 2$ can begin with its first column covered by two squares, two horizontal dominoes, or a square either over or under a domino. The remaining part of the board is a rectangular board of length $n + 1$, or a rectangular board of length n, or a pruned board of length $n + 1$with a missing upper or lower corner cell. Thus, $u_{n+2} = u_{n+1} + u_n + 2v_{n+1}$. Similarly, a pruned board of length $n + 1$ has the single cell in its first column covered either by a square or domino, showing that $v_{n+1} = u_n + v_n$.

(b) In terms of the E operator, the recurrence relations have the form $\left(E^2 - E - 1\right) u_n - 2Ev_n = 0$ and $-u_n + (E - 1)\,v_n = 0$. By the result of Problem 5.4.15, the operator $T(E) = \left(E^2 - E - 1\right)(E - 1) - 2E =$

$E^3 - 2E^2 - 2E + 1$ annihilates both u_n and v_n. That is, both sequences satisfy the third order recurrence relation $h_{n+3} = 2h_{n+2} + 2h_{n+1} - h_n$.

(c) A $1 \times n$ board can be tiled with squares and (necessarily) horizontal dominoes in $f_n = F_{n+1}$ ways. Since the horizontal domino restriction means both rows of either a complete or a pruned $2 \times n$ board are tiled independently of each other, it follows that that $u_n = F_{n+1}^2$ and $v_n = F_n F_{n+1}$.

5.4.19. Find a fourth-order recurrence relation satisfied by F_n^3, L_n^3,$F_n^2 L_n$, and $F_n L_n^2$.

Answer

By the Binet formulas $L_n = \varphi^n + \hat{\varphi}^n$ and $F_n = \dfrac{\varphi^n - \hat{\varphi}^n}{\sqrt{5}}$, each of F_n^3, L_n^3, $F_n^2 L_n$, and $F_n L_n^2$ is a GPS with the eigenvalues $\{\varphi^3, \hat{\varphi}^3, -\varphi, -\hat{\varphi}\}$. For example,

$$\begin{aligned} L_n^3 &= (\varphi^n + \hat{\varphi}^n)^3 = \varphi^{3n} + 3\varphi^{2n}\hat{\varphi}^n + 3\varphi^n\hat{\varphi}^{2n} + \hat{\varphi}^{3n} \\ &= \varphi^{3n} + 3\,(-\varphi)^n + 3\,(-\hat{\varphi})^n + \hat{\varphi}^{3n}. \end{aligned}$$

Therefore, F_n^3, L_n^3, $F_n^2 L_n$, and $F_n L_n^2$ are each annihilated by

$$\begin{aligned} &\left(E - \varphi^3\right)\left(E - \hat{\varphi}^3\right)(E + \varphi)(E + \hat{\varphi}) = \left(E^2 - L_3 E - 1\right)\left(E^2 + L_1 E - 1\right) \\ &= \left(E^2 - 4E - 1\right)\left(E^2 + E - 1\right) = E^4 - 3E^3 - 6E^2 + 3E + 1 \end{aligned}$$

so each sequence satisfies the same fourth-order recurrence relation

$$h_{n+4} - 3h_{n+3} - 6h_{n+2} + 3h_{n+1} + h_n = 0.$$

PROBLEM SET 5.5

5.5.1. Consider the nonhomogeneous recurrence relation $h_{n+2} = h_{n+1} + 6h_n + 6$ with the initial conditions $h_0 = 0, h_1 = 2$.

(a) Determine a homogeneous recurrence relation of order 3 satisfied by h_n.

(b) Solve the recurrence relation.

(c) Is there a homogeneous relation of order 2 satisfied by h_n?

Answer

(a) The associated homogeneous recurrence is annihilated by $C(E) = E^2 - E - 6 = (E - 3)(E + 2)$ and the nonhomogeneous term $q_n = 6$

is annihilated by $P(E) = E - 1$. Therefore, h_n is annihilated by the operator $(E - 3)(E + 2)(E - 1)$.

(b) By part (a), h_n is a GPS of the three distinct eigenvalues 3, –2, and 1. That is, $h_n = c_1 3^n + c_2 (-2)^n + c_3$. Using the additional initial condition $h_2 = h_1 + 6h_0 + 6 = 2 + 0 + 6 = 8$ we get $h_n = 3^n - 1$.

(c) The solution $h_n = 3^n - 1$ is annihilated by $(E - 3)(E - 1) = E^2 - 4E + 3$ so h_n also satisfies $h_{n+2} = 4h_{n+1} - 3h_n$.

5.5.3. Solve these nonhomogeneous recurrence relations.

(a) $h_{n+1} = 2h_n + 3^n + 1, h_0 = 0$

(b) $h_{n+1} = 2h_n + n3^n, h_0 = 0$

Answer

(a) The associated homogeneous recurrence relation is annihilated by $E - 2$, $q_n = 3^n$ is annihilated by $E - 3$, and $\hat{q}_n = 1$ is annihilated by $E - 1$. Thus, h_n is annihilated by $(E - 2)(E - 3)(E - 1)$, so h_n has the form $h_n = c_1 2^n + c_2 3^n + c_3$. The initial condition, with $h_1 = 2h_0 + 3^0 + 1 = 2$ and $h_2 = 2h_1 + 3^1 + 1 = 8$ gives the solution $h_n = 3^n - 1$.

(b) The associated homogeneous recurrence relation is annihilated by $E - 2$ and $q_n = n3^n$ is annihilated by $(E - 3)^2$. Therefore, h_n is annihilated by $(E - 2)(E - 3)^2$, so h_n has the form $h_n = c_1 2^n + c_2 3^n + c_3 n3^n$. The initial condition, together with $h_1 = 2h_0 + 0 \cdot 3^0 = 0$ and $h_2 = 2h_1 + 1 \cdot 3^1 = 3$, gives the solution $h_n = 3 \cdot 2^n - 3 \cdot 3^n + n3^n$.

5.5.5. One line separates the plane into $h_1 = 2$ regions, and two intersecting lines separate the plane into $h_2 = 4$ regions. Three lines, none parallel and no three intersecting at a single point, separate the plane into $h_3 = 7$ regions:

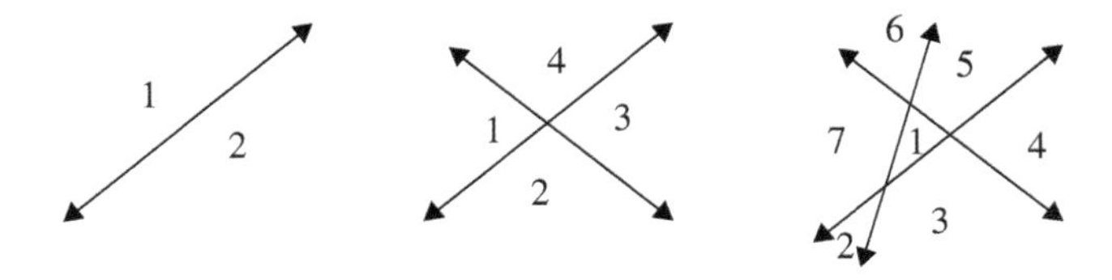

Now let h_n denote the number of regions determined by n lines in the plane, where no two lines are parallel and no three lines intersect at the same point.

(a) Explain why h_n satisfies the nonhomogeneous recurrence relation $h_{n+1} = h_n + n + 1$.

[*Hint*: How many additional regions are created as the $(n + 1)$st line is drawn?]

(b) Solve the homogeneous recurrence relation $h_{n+1} = h_n$.

(c) Show there is a particular solution of the nonhomogeneous of the form $p_n = sn^2 + tn$.

(d) Determine h_n.

Answer

(a) Each intersection of the $(n+1)^{\text{st}}$ line with one of the existing n lines creates a new region, and another region is created beyond the last intersection. Thus, $n+1$ new regions are formed by the $(n+1)^{\text{st}}$ line.

(b) The characteristic equation of the associated homogeneous recurrence relation is $x - 1 = 0$, so the eigenvalue is $\alpha = 1$ and the general solution of the homogeneous equation is a constant: $g_n = c$.

(c) The nonhomogeneous term $q_n = n + 1$ is a linear polynomial in the variable n. However, constants solve the associated homogeneous recurrence relation, so we look for a particular solution of the form $p_n = sn^2 + tn$. When inserted into the nonhomogeneous recurrence, we get $p_{n+1} = s(n+1)^2 + t(n+1) = p_n + n + 1 = sn^2 + tn + n + 1$ which simplifies to become $(2s-1)n + (s+t-1) = 0$. This holds for all $n \geq 0$ if $s = t = 1/2$, giving us the particular solution $p_n = \frac{1}{2}n(n+1)$. Therefore, $h_n = g_n + p_n = c + \frac{1}{2}n(n+1)$. Since $h_0 = 1$ (no lines correspond to 1 region, namely the entire plane), we see that $c = 1$, and the final solution is $h_n = 1 + \frac{1}{2}n(n+1)$.

5.5.7. The sequence $p_n^{(r)}$ of r-gonal numbers is obtained by starting with a single vertex point, so $p_1^{(r)} = 1$ and then successively adding additional points so the outermost polygon has $n + 1$ dots along each of its r sides. For example, the hexagonal numbers are shown in the diagram below. Note that $p_0^{(r)} = 0, p_1^{(r)} = 1$, and $p_2^{(r)} = r$ for all $r \geq 3$.

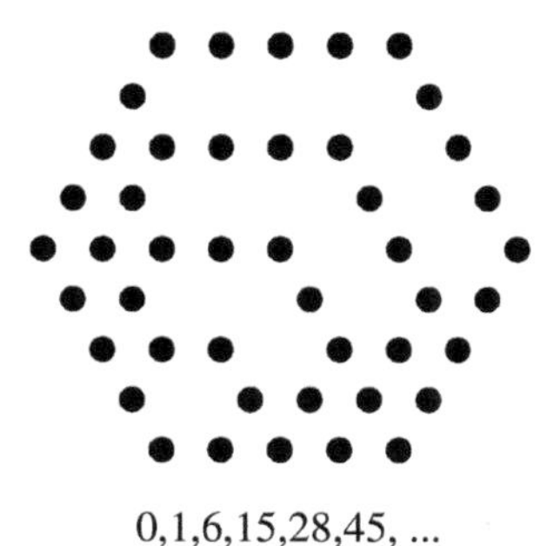

0,1,6,15,28,45, ...

(a) Show that $p_{n+1}^{(r)} = p_n^{(r)} + 1 + (r-2)\,n$.

(b) Explain why the sequence of r-gonal numbers is annihilated by the operator $C(E) = (E-1)^3$.

(c) Solve the recurrence to show that $p_n^{(r)} = n + (r-2)\binom{n}{2}$.

Answer

(a) The $(n+1)$st polygon adds $1+(r-2)\,n$ dots to the number of dots given by $p_n^{(r)}$.

(b) The associated homogeneous recurrence $g_n = g_{n-1}$ is annihilated by $E-1$ and the nonhomogeneous term $q_n = 1+(r-2)\,n$ is a first degree polynomial annihilated by $(E-1)^2$. Therefore the sequence of r-gonal numbers is annihilated by $(E-1)(E-1)^2 = (E-1)^3$.

(c) By part (b), $p_n^{(r)}$ is a quadratic polynomial $p_n^{(r)} = c_0 + c_1\binom{n}{1} + c_2\binom{n}{2}$. Using the initial conditions $p_0^{(r)} = 0, p_1^{(r)} = 1$, and $p_2^{(r)} = r$ we find that $c_0 = 0, c_1 = 1$, and $c_2 = r-2$ so $p_n^{(r)} = n + (r-2)\binom{n}{2}$. Often this formula is expressed in the form $p_n^{(r)} = \frac{n}{2}\left[(n-1)r - 2(n-2)\right]$ or $p_n^{(r)} = \frac{n}{2}\left[(r-2)n - (r-4)\right]$.

5.5.9. The number of ways to tile an n-board with red and blue squares and white dominoes is given by the Pell number P_{n+1} (see Problem 5.3.15). Let r_n and d_n denote, respectively, the number of red squares and the number of dominoes in *all* of the tilings of boards of length n. By symmetry, r_n also is the number of blue squares in all of the tilings.

(a) Create a table analogous to the table shown in Example 5.35, but for Pell tilings of length $n = 0, 1, 2$, and 3. Verify that $r_1 = 2, d_1 = 1, r_2 = 4, d_2 = 1$, and $r_3 = 14, d_3 = 4$.

(b) Find a recurrence relation satisfied by r_n.

(c) What is the form of the GPS for r_n, with unspecified coefficients?

(d) Verify that $r_n = \frac{1}{4}\left((n+1)\,P_n + nP_{n+1}\right)$.

(e) Prove that $d_n = \frac{1}{4}\left(nP_{n-1} + (n-1)\,P_n\right)$.

(f) Prove that $r_n = d_{n+1}$.

Answer

(a)

n	All tilings of a $1 \times n$ board with red and blue squares and white dominoes [white domino], red square = [red square], blue square = [blue square]	Number of tilings P_{n+1}	Number of red squares r_n	Number of white dominoes d_n
0		1	0	0
1		2	1	0
2		5	4	1
3	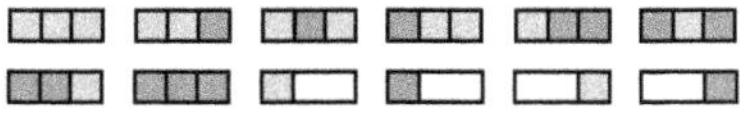	12	14	4

(b) The P_{n+1} boards of length n are of three types depending on the type of the first tile. There are P_n boards that begin with a red square and these account for $P_n + r_{n-1}$ red squares. There are also r_{n-1} red squares used in the tilings that begin with a blue square and r_{n-2} red squares in the tilings that begin with a white domino. Altogether, we see that

$$r_n = \left(P_n + r_{n-1}\right) + r_{n-1} + r_{n-2} = 2r_{n-1} + r_{n-2} + P_n.$$

(c) Both the associated homogeneous recurrence relation and the nonhomogeneous term $q_n = P_n$ are annihilated by the Pell operator $C(E) = E^2 - 2E - 1$, so we know that r_n is a GPS of the eigenvalues of the equation $\left(x^2 - 2x - 1\right)^2 = 0$. That is, r_n has the form

$$r_n = c_1\left(1+\sqrt{2}\right)^n + c_2\left(1-\sqrt{2}\right)^n + c_3 n\left(1+\sqrt{2}\right)^n + c_4 n\left(1-\sqrt{2}\right)^n.$$

(d) We can also write, although with different constants, that $r_n = c_1P_n + c_2P_{n+1} + c_3nP_n + c_4nP_{n+1}$. From the recurrence and part (a) we have the additional initial condition $r_0 = 0$. The formula $r_n = \frac{1}{4}\left(P_n + nP_n + nP_{n+1}\right)$ can then be checked to have the four required initial conditions. Indeed, $r_0 = \frac{1}{4}\left(P_0 + 0 \cdot P_0 + 0 \cdot P_1\right) = 0, r_1 = \frac{1}{4}\left(P_1 + P_1 + P_2\right) = \frac{1}{4}(1+1+2) = 1, r_2 = \frac{1}{4}\left(P_2 + 2P_2 + 2P_3\right) =$

$\frac{1}{4}(2+2\cdot 2+2\cdot 5)=\frac{16}{4}=4,$ and $r_3=\frac{1}{4}\left(P_3+3P_3+3P_4\right)=\frac{1}{4}(5+3\cdot 5+3\cdot 12)=\frac{56}{4}=14.$

(e) The P_{n+1} boards of length n cover an area of nP_{n+1}. This area is covered by $2r_n=\frac{1}{2}\left((n+1)P_n+nP_{n+1}\right)$ red and blue squares and the d_n dominoes of area $2d_n$. Thus, $nP_{n+1}=\frac{1}{2}\left((n+1)P_n+nP_{n+1}\right)+2d_n$. Solving for d_n gives $d_n=\frac{1}{4}\left((n-1)P_n+nP_{n-1}\right)$.

(f) $d_{n+1}=\frac{1}{4}\left(nP_{n+1}+(n+1)P_n\right)=r_n$.

5.5.11. Suppose a $2\times n$ board is tiled with gray and white squares, where in each row the white squares are to the left of the gray squares. Three examples of tilings of a 2×9 board are shown here.

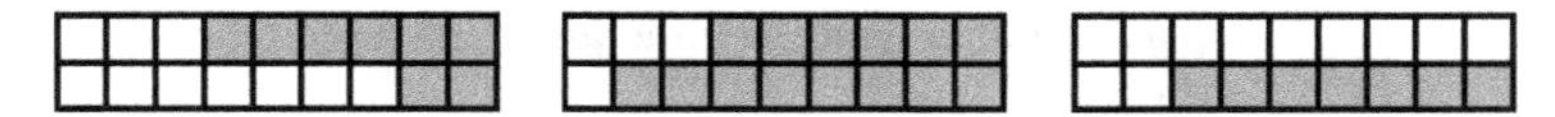

Determine the number of tilings by

(a) finding and solving a recurrence relation

(b) using direct combinatorial reasoning

Answer

(a) Let h_n denote the number of tilings of a board of length n. Then $h_1=4$, since each cell can be either white or gray. Any tiling of a $2\times n$ board for $n\geq 2$ has h_{n-1} ways to begin with white square covering each cell of the first column. If the first column in covered with a white and a gray square, the tiling is completed in n ways by giving the number, $0,1,\ldots,n-1$, of additional white squares in the row that begins with a white square. Finally, there is the tiling with all gray squares. Thus, $h_n=h_{n-1}+2n+1,\ n\geq 2$, which holds for $n=1$ if we set $h_0=1$. We also see that $h_2=h_1+2\cdot 2+1=9$. The associated homogeneous recurrence is annihilated by $E-1$ and the nonhomogeneous term $q_n=2n+1$ is annihilated by $(E-1)^2$. This means $\alpha=1$ is an eigenvalue of mulitiplicity 3 and thus h_n is a GPS of the form $c_0+c_1n+\binom{n}{2}c_2$. Then $h_0=1=c_0$ and $h_1=1+c_1=4$, so $c_0=1$ and $c_1=3$. Then substituting into the recurrence shows $9=h_2=1+3\cdot 2+c_2\binom{2}{2}$ so $c_2=2$. Therefore, $h_n=1+3n+n(n-1)=(n+1)^2$.

(b) In each row, there are $n+1$ choices for the number of white squares that are to the left of the row. This gives $(n+1)^2$ tilings.

5.5.13. Consider all of the P_n tilings of boards of length n with red and blue squares and white dominoes, where P_n is the nth Pell number. In Problem 5.5.9, it was shown that the number of red squares, r_n, used to tile the n-boards was the same as the number of white dominoes, d_{n+1}, used to create all of the tilings of boards of length $n+1$. For example, $r_2 = 4 = d_3$ as shown in the following diagram.

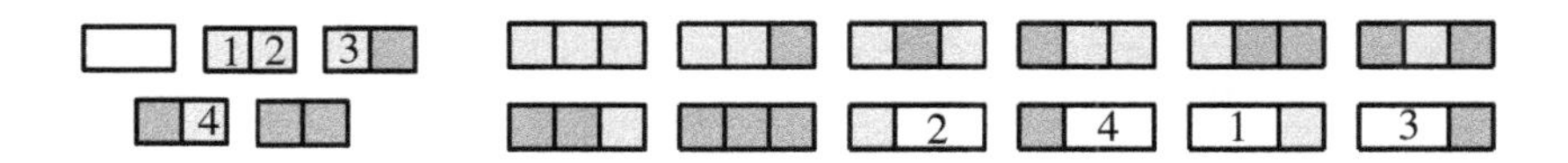

Give a bijective proof that $r_n = d_{n+1}$.

Answer
Number the red squares 1, 2, …, r_n used in the tilings of n-boards. If square j is replaced with a white domino also numbered by j, this creates a tiling of a board of length n +1. The mapping between red squares and dominoes is a bijection since it is invertible: any domino of a tiling of a board of length $n+1$ can be replaced by a red square to leave a tiling of an n-board with a red square that was assigned a number from 1 to r_n.

PROBLEM SET 5.6

5.6.1. The sequence that gives the number of ways to tile a $1 \times n$ board with red squares and blue or green dominoes satisfies the recurrence relation $h_n = h_{n-1} + 2h_{n-2}$ with the initial conditions $h_0 = 1$ and $h_1 = 1$. Determine the ordinary generating function for this sequence.

Answer
The sequence is annihilated by the operator $C(E) = E^2 - E - 2$, so the OGF has the form $f(x) = 1 + x + \cdots = \dfrac{b_0 + b_1 x}{1 - x - 2x^2}$. Therefore, $\left(1 - x - 2x^2\right) f(x) = \left(1 - x - 2x^2\right)(1 + x + \cdots) = 1 + 0x + \cdots = b_0 + b_1 x$ so $b_0 = 1$ and $b_1 = 0$, giving the OGF $f(x) = \dfrac{1}{1 - x - 2x^2}$.

5.6.3. If h_n denotes the number of ways to position n nonattacking kings on a $2 \times 2n$ board, then it has been shown that $h_n = 4h_{n-1} - 4h_{n-2}$, $n \geq 2$, where $h_0 = 1$ and $h_1 = 4$. Show that the OGF of the sequence h_n is given by $f(x) = \dfrac{1}{1 - 4x + 4x^2}$.

Answer

Since the sequence in annihilated by $C(E)=E^2-4E+4$, the OGF has the form $f(x)=1+4x+\cdots=\dfrac{b_0+b_1x}{1-4x+4x^2}$. Therefore, $\left(1-4x+4x^2\right)f(x)=\left(1-4x+4x^2\right)(1+4x+\cdots)=1-0x+\cdots=b_0+b_1x$ so $b_0=1$ and $b_1=0$, giving the OGF $f(x)=\dfrac{1}{1-4x+4x^2}$.

5.6.5. The number of ways that a $2\times n$ board can be tiled with squares and dominoes is given by the sequence u_n that satisfies the recurrence relation $u_{n+3}=3u_{n+2}+u_{n+1}-u_n$, with the initial conditions $u_0=1, u_1=2, u_2=7$. Obtain the OGF for this sequence.

Answer

Since the characteristic polynomial is $C(x)=x^3-3x^2-x+1$, the OGF has the form $f(x)=1+2x+7x^2+\cdots=\dfrac{b_0+b_1x+b_2x^2}{1-3x-x^2+x^3}$. Therefore, $\left(1-3x-x^2+x^3\right)f(x)=\left(1-3x-x^2+x^3\right)\left(1+2x+7x^2+\cdots\right)=1-x+0x^2+\cdots=b_0+b_1x+b_2x^2$ so $b_0=1, b_1=-1$, and $b_2=0$, giving the OGF $f(x)=\dfrac{1-x}{1-3x-x^2+x^3}$.

5.6.7. Find the OGF of the sequence given by the nonhomogeneous recurrence relation $h_{n+2}=4h_{n+1}-4h_n+5\cdot 2^n, n\geq 0$ with the initial conditions $h_0=0, h_1=2$.

Answer

The associated homogeneous recurrence relation is annihilated by the operator E^2-4E+4 and the nonhomogeneous term is annihilated by the operator $E-2$, so the sequence h_n is annihilated by $\left(E^2-4E+4\right)(E-2)=E^3-6E^2+12E-8$. Thus the OGF has the form $f(x)=0+2x+\cdots=\dfrac{b_0+b_1x+b_2x^2}{1-6x+12x^2-8x^3}$. From the initial conditions we calculate that $h_2=4\cdot 2-4\cdot 0+5\cdot 2^0=13$. Therefore,

$$
\begin{aligned}
\left(1-6x+12x^2-8x^3\right)f(x)&=\left(1-6x+12x^2-8x^3\right)\left(0+2x+13x^2+\cdots\right)\\
&=0+2x+x^2+\cdots=b_0+b_1x+b_2x^2
\end{aligned}
$$

so $b_0=0$, $b_1=2$, and $b_2=1$, giving the OGF $f(x)=\dfrac{2x+x^2}{1-6x+12x^2-8x^3}$.

5.6.9. Recall that the sequence $p_n^{(r)}$ of r-gonal numbers satisfies the recurrence relation $p_{n+1}^{(r)}=p_n^{(r)}+1+(r-2)n$ with the initial conditions $p_0^{(r)}=$

$0, p_1^{(r)} = 1$, and $p_2^{(r)} = r$ (see Problem 5.5.7). Show that the ordinary generating function $f^{(r)}$ for the r-gonal numbers is $f^{(r)}(x) = \dfrac{x + (r-3)x^2}{(1-x)^3}$.

Answer
Since $p_0^{(r)} = 0$, there is a quadratic polynomial $P(x) = b_1 x + b_2 x^2$ for which

$$f^{(r)}(x) = x + rx^2 + \cdots = \frac{b_1 x + b_2 x^2}{(1-x)^3}.$$

Thus $b_1 x + b_2 x^2 = (1 - 3x + \cdots)\left(x + rx^2 + \cdots\right) = x + (r-3)x^2 + \cdots$ so $b_1 = 1$ and $b_2 = r - 3$. This gives the desired OGF.

5.6.11. Obtain the OGF f_F of the Fibonacci numbers by working with the three series

$$\begin{aligned} f_F(x) &= F_0 + F_1 x + F_2 x^2 + F_3 x^3 + F_4 x^4 + F_5 x^5 + \cdots \\ x f_F(x) &= \phantom{F_0 + {}} F_0 x + F_1 x^2 + F_2 x^3 + F_3 x^4 + F_4 x^5 + \cdots \\ x^2 f_F(x) &= \phantom{F_0 + F_1 x + {}} F_0 x^2 + F_1 x^3 + F_2 x^4 + F_3 x^5 + \cdots . \end{aligned}$$

In particular, subtract the second and third series from the first.

Answer
Since $F_{n+2} = F_{n+1} + F_n$, subtracting the second and third series from the first shows that $\left(1 - x - x^2\right) f_F(x) = F_0 + \left(F_1 - F_0\right) x = 0 + (1-0)x = x$. Solving the equation for f_F gives $f_F(x) = \dfrac{x}{1 - x - x^2}$.

5.6.13. Obtain the OGF of the (a) Perrin and (b) tribonacci numbers by modifying the idea of Problem 5.6.11.

Answer

(a)

$$\begin{aligned} f_P(x) &= p_0 + p_1 x + p_2 x^2 + p_3 x^3 + p_4 x^4 + \cdots \\ x^2 f_p(x) &= \phantom{p_0 + p_1 x + {}} p_0 x^2 + p_1 x^3 + p_2 x^3 + \cdots \\ x^3 f_p(x) &= \phantom{p_0 + p_1 x + p_2 x^2 + {}} p_0 x^3 + p_1 x^3 + \cdots \\ \hline \left(1 - x^2 - x^3\right) f_p(x) &= p_0 + p_1 x + (p_2 - p_0) x^2 = 3 + (2-3)x^2 = 3 - x^2 \end{aligned}$$

Thus $f_p(x) = \dfrac{3 - x^2}{1 - x^2 - x^3}$.

(b) By a calculation similar to that of part (a), $\left(1 - x - x^2 - x^3\right) f_T(x) = T_0 + \left(T_1 - T_0\right) x + \left(T_2 - T_1 - T_0\right) x^2 = 0 + (0-0)x + (1-0-0)$ $x^2 = x^2$

Thus, $f_T(x) = \dfrac{x^2}{1 - x - x^2 - x^3}$.

5.6.15. Obtain the EGF of the Fibonacci sequence, using the Binet formula $F_n = \dfrac{\varphi^n - \hat{\varphi}^n}{\sqrt{5}}$.

Answer

$$f_F^{(e)}(x) = \sum_{n=0}^{\infty} F_n \frac{x^n}{n!} = \sum_{n=0}^{\infty} \left(\frac{\varphi^n - \hat{\varphi}^n}{\sqrt{5}} \right) \frac{x^n}{n!} = \frac{1}{\sqrt{5}} \left(e^{\varphi x} - e^{\hat{\varphi} x} \right).$$

5.6.17. Suppose there are m students $s_1, s_2, \ldots, s_m$ and n tutors $t_1, t_2, \ldots, t_n$. Let $\langle m, n \rangle$ denote the number a ways the students can choose which tutor, if any, with whom to work. This can be illustrated with a diagram known as a *bipartite graph* as shown below for the case $m = 4$ and $n = 5$.

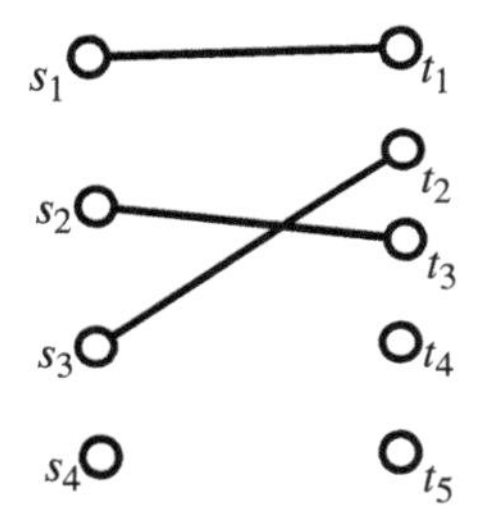

Note that some students do not choose a tutor and some tutors have no students. However, no student has more than one tutor, and no tutor has more than one student. Such a pairing is called a *matching* in graph theory.

(a) Explain why $\langle m, 1 \rangle = m + 1$, $\langle 1, n \rangle = n + 1$ and $\langle m, n \rangle = \langle n, m \rangle$.

(b) Derive the recurrence relation $\langle m, n \rangle = \langle m, n-1 \rangle + m \langle m-1, n-1 \rangle$.

(c) Explain why $\langle m, n \rangle = \sum_{k \geq 0} (m)_k \binom{n}{k}$. (Recall that $(m)_k$ is the number of k-permutations of an m set.)

(d) Show that $f_m^{(e)}(y) = \sum_{n=0}^{\infty} \langle m, n \rangle \dfrac{y^n}{n!} = (1 + y)^m e^y$, where $\langle m, 0 \rangle = 1$ and $\langle 0, n \rangle = 1$.

(e) Show that $f^{(e)}(x, y) = \sum_{m,n=0}^{\infty} \langle m, n \rangle \dfrac{x^m y^n}{m! n!} = e^{x+y+xy}$.

Answer

(a) The one tutor can work with any of the m students or no student, so $\langle m, 1 \rangle = m + 1$. Similarly, the one student can work with any one of the n tutors or none at all so $\langle 1, n\rangle = n+1$. Since the role of tutor and student can be reversed, $\langle m, n\rangle = \langle n, m\rangle$.

(b) Take cases depending on whether tutor t_n works with a student or not. If not, there are $\langle m, n-1\rangle$ student-tutor pairings. Otherwise tutor t_n can work with any of m students and the unassigned students and tutors can be paired in $\langle m-1, n-1\rangle$ ways.

(c) Let k be the number of students to be assigned a tutor, so there are $\binom{n}{k}$ tutors that can be paired with k students in $(n)_k$ ways. Summing over k gives all pairings.

(d) $$f_m^{(e)}(y) = \sum_{n\geq 0} \langle m, n\rangle \frac{y^n}{n!} = \sum_{n\geq 0}\sum_{k\geq 0} (m)_k \binom{n}{k} \frac{y^n}{n!}$$

$$= \sum_{k\geq 0}\left((m)_k \frac{y^k}{k!} \sum_{n\geq k} \frac{y^{n-k}}{(n-k)!}\right) = \sum_{k\geq 0} \binom{m}{k} y^k e^y = (1+y)^m e^y$$

(e) $$f^{(e)}(x, y) = \sum_{m,n=0}^{\infty} \langle m, n\rangle \frac{x^m y^n}{m!n!} = \sum_{m\geq 0} (1+y)^m \frac{x^m}{m!} e^y = e^{x(1+y)} e^y$$

$$= e^{x+y+xy}$$

5.6.19. Let $f^{(e)}(x) = \sum_{m=0}^{\infty} \frac{x^{3m}}{(3m)!}$ be the EGF of the sequence g_n satisfying $g_{n+3} = g_n$, $g_0 = 1, g_1 = 0$, and $g_2 = 0$. Show that

$$f^{(e)}(x) = \frac{1}{3}(e^x + e^{\omega x} + e^{\omega^2 x}) = \frac{1}{3}\left(e^x + 2e^{-x/2}\cos\left(\frac{\sqrt{3}}{2}x\right)\right),$$

where ω is the primitive third root of unity $\omega = e^{2\pi i/3} = -\frac{1}{2} + i\frac{\sqrt{3}}{2}$. That is, $\omega^3 = 1$ and $1 + \omega + \omega^2 = 0$.

Answer

Since $C(x) = x^3 - 1 = (x-1)(x-\omega)(x-\omega^2)$, the eigenvalues are $1, \omega$, and ω^2. Therefore the EGF has the form

$$f^{(e)}(x) = \sum_{n=0}^{\infty} \left(c_0 + c_1\omega^n + c_2\omega^{2n}\right)\frac{x^n}{n!} = c_0 e^x + c_1 e^{\omega x} + c_2 e^{\omega^2 x}$$

$$= \left(c_0 + c_1 + c_2\right) + \left(c_0 + c_1\omega + c_2\omega^2\right)x$$

$$+ \left(c_0 + c_1\omega^2 + c_2\omega^4\right)\frac{x^2}{2!} + \cdots.$$

It is easy to check that $c_0 = c_1 = c_2 = \dfrac{1}{3}$:

$$\frac{1}{3} + \frac{1}{3} + \frac{1}{3} = 1, \frac{1}{3} + \frac{1}{3}\omega + \frac{1}{3}\omega^2 = \frac{1}{3}(1 + \omega + \omega^2) = 0, \frac{1}{3} + \frac{1}{3}\omega^2 + \frac{1}{3}\omega^4$$
$$= \frac{1}{3} + \frac{1}{3}\omega^2 + \frac{1}{3}\omega = 0$$

Therefore,

$$f^{(e)}(x) = \frac{1}{3}\left(e^x + e^{\omega x} + e^{\omega^2 x}\right) = \frac{1}{3}\left(e^x + 2\mathrm{Re}\left(e^{\omega x}\right)\right)$$
$$= \frac{1}{3}\left(e^x + 2\mathrm{Re}e^{\left(-\frac{1}{2}+i\frac{\sqrt{3}}{2}\right)x}\right) = \frac{1}{3}\left(e^x + 2e^{-x/2}\cos\left(\frac{\sqrt{3}}{2}x\right)\right).$$

5.6.21. Let $f(x) = \sum_{n\geq 0} a_n x^n$ be the OGF of the sequence a_n for which $na_n = a_{n-1} + a_{n-2}, n \geq 2$ and $a_0 = 1, a_1 = 1$.

(a) Show that $f'(x) = (1 + x)f(x)$.

(b) Solve the differential equation in part (a) to show that $f(x) = e^{x+x^2/2}$.

Answer

(a) $f'(x) = \sum_{n\geq 1} na_n x^{n-1} = a_1 + \sum_{n\geq 2} na_n x^{n-1} = 1 + \sum_{n\geq 2} a_{n-1}x^{n-1} + x\sum_{n\geq 2} a_{n-2}x^{n-2} = f(x) + xf(x)$

(b) $\dfrac{f'(x)}{f(x)} = 1 + x$ so $\log f(x) = c + x + \dfrac{x^2}{2}$. Then $\log f(0) = \log 1 = 0 = c$ so $f(x) = e^{x+\frac{x^2}{2}}$.

PROBLEM SET 5.7

5.7.1. Use a piece of $8'' \times 8''$ squared paper to cut out the four shapes shown here.

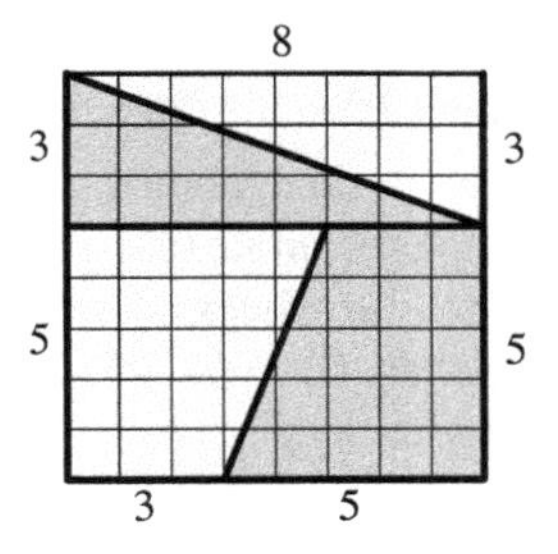

(a) Rearrange the four pieces to form this $5'' \times 13''$ rectangle. How does the area of the rectangle compare to the area of the square?

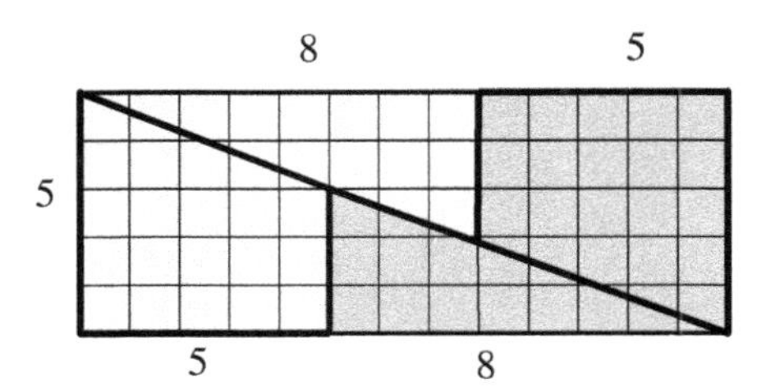

(b) Rearrange the 4 pieces to form this "propeller" shape. What is its area?

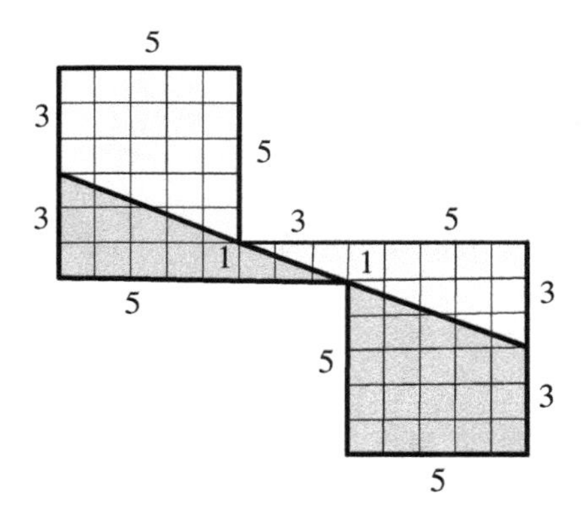

(c) Explain the excess or missing areas.

Answer

(a) The 5 by 13 rectangle has area 65 square inches, compared to the area 64 square inches of the square.

(b) The propeller has area $30 + 3 + 30 = 63$ square inches.

(c) The four pieces leave a thin rhombus of area 1 square inch running diagonally through the rectangle that is not covered by the four pieces: $5 \times 13 = 8^2 + 1$. The four pieces actually overlap by 1 square inch as they cover the propeller: $63 = 64 - 1$. These geometric puzzles are a consequence of the identity $F_{n+1}F_{n-1} - F_n^2 = (-1)^n$.

5.7.3. The following diagram describes a mapping that pairs the set $S = \{1, 4, 6, 9\} \subseteq [9]$ with the block walking path ENEENE.

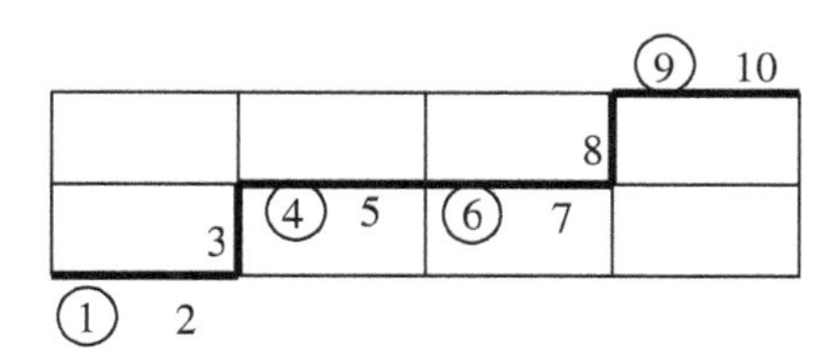

Generalize this bijection to show there are $\binom{n+1-k}{k}$ subsets of $[n]$ with k elements, with no two consecutive elements.

Answer

There are $\binom{n+1-k}{k}$ paths from $A(0, 0)$ to the point $B(k, n+1-2k)$ that cross k east blocks and $n + 1 - 2k$ north blocks. Each path can be numbered by the set $[n]$ with one number assigned to each north block and two to each east block. The first numbers s_i assigned to the k horizontal easterly blocks along the path define a subset $\{s_1, s_2, \ldots, s_k\} \subseteq [n]$ of k elements with no two consecutive elements. Moreover, the mapping is invertible, since each set S defines a unique block walk.

5.7.5. **(a)** Use mathematical induction to prove that any pair of successive Fibonacci numbers F_m and $F_{m+1}, m \geq 0$ have no common integer divisor larger that 1. That is, show that $\gcd(F_m, F_{m+1}) = 1$, where gcd is the *greatest common divisor*. Note that $\gcd(a, b) = \gcd(a, ka + r) = \gcd(a, r)$ if $a = kb + r, 0 \leq r < a$.

(b) Let $0 < m < n$, so $n = km + r$ for some $0 \leq r < m$. Prove that $\gcd(F_m, F_n) = \gcd(F_m, F_r)$ and $\gcd(m, n) = \gcd(m, r)$. [*Hint*: In Problem 5.2.3, it was shown that $F_{km+r} = F_r F_{km+1} + F_{r-1} F_{km}$ and F_k divides F_{mk}.]

(c) Use part (b) to prove that $\gcd(F_m, F_n) = F_{\gcd(m,n)}$. [*Hint*: Assume that the theorem is incorrect, and let m the smallest positive integer for which the formula doesn't hold for some pair of Fibonacci numbers F_m and F_n.]

Answer

(a) First note that $\gcd(F_0, F_1) = \gcd(0, 1) = 1$. Now let $m > 0$ and suppose that $\gcd(F_m, F_{m+1}) = 1$. Then $\gcd(F_{m+1}, F_{m+2}) = \gcd(F_{m+1}, F_m + F_{m+1}) = \gcd(F_{m+1}, F_m) = 1$, completing the induction.

(b) Using the hint we see that

$$\begin{aligned}\gcd\left(F_m, F_n\right) &= \gcd\left(F_m, F_{km+r}\right) = \gcd\left(F_m, F_r F_{km+1} + F_{r-1} F_{km}\right)\\ &= \gcd\left(F_m, F_r F_{km+1}\right) \quad \left[\text{since } F_m \text{ divides } F_{km}\right]\\ &= \gcd\left(F_m, F_r\right) \quad \left[\begin{array}{c}\text{since any integer greater than 1 that}\\ \text{divides } F_m \text{ also divides } F_{km} \text{ and}\\ F_{km} \text{ and } F_{km+1} \text{ have no common}\\ \text{divisor except 1}\end{array}\right]\end{aligned}$$

so $\gcd\left(F_m, F_n\right) = \gcd\left(F_m, F_r\right)$. Also $\gcd(m, n) = \gcd(m, km + r) = \gcd(m, r)$, proving that $\gcd(m, n) = \gcd(m, r)$.

(c) Assume that $\gcd\left(F_m, F_n\right) \neq F_{\gcd(m,n)}$ where $m < n$ and m is the smallest positive integer for which this is so. But by part (b), if $n = km + r$ with $r < m$, we have shown that $\gcd\left(F_m, F_r\right) = \gcd\left(F_m, F_n\right) \neq F_{\gcd(m,n)} = F_{\gcd(m,r)}$. This is a contradiction to the choice of a smallest m.

5.7.7. The *golden ratio* $\varphi = \left(1 + \sqrt{5}\right)/2$ originated in ancient Greek mathematics with this problem. How can a line segment AB be divided into two segments AP and PB so that $\dfrac{AB}{AP} = \dfrac{AP}{PB}$?

Prove that the common ratio is given by the golden ratio.

Answer

Let $AP = xPB$. Since $AB = AP + PB = (x + 1)\,PB$, the proportion becomes $\dfrac{(x+1)\,PB}{xPB} = \dfrac{xPB}{PB}$, so $x + 1 = x^2$. The unique positive solution is $x = \dfrac{1+\sqrt{5}}{2} = \varphi$.

5.7.9. Prove that $F_n L_{n+1} - F_{n+1} L_n = (-1)^{n+1}\,2$ for all $n \geq 0$.

Answer

The formula holds for $n = 0$ since $0{\cdot}1 - 1{\cdot}2 = -2$. Suppose the formula holds for n, where $n \geq 0$. Then

$$\begin{aligned}F_{n+1} L_{n+2} - F_{n+2} L_{n+1} &= F_{n+1}\left(L_{n+1} + L_n\right) - \left(F_{n+1} + F_n\right) L_{n+1}\\ &= F_{n+1} L_n - F_n L_{n+1} = -(-1)^{n+1}\,2 = (-1)^{n+2}\,2\end{aligned}$$

so the formula holds for all $n \geq 0$ by mathematical induction.

5.7.11. Let $C(E)$ annihilate the sequence u_n and let $D(E)$ annihilate the sequence v_n. Determine annihilating operators of the sequences

(a) $u_n + v_n$.

(b) $w_n = \sum_{k=0}^{n} u_k$.

Answer

(a) $C(E)D(E)\left(u_n + v_n\right) = D(E)C(E)u_n + C(E)D(E)v_n = 0 + 0 = 0$

(b) $C(E)(E-1)w_n = C(E)\left(\sum_{k=0}^{n+1} u_k - \sum_{k=0}^{n} u_k\right) = C(E)u_{n+1} = 0$

5.7.13. **(a)** Show that the Pell equation $x^2 - 5y^2 = -4$ is solved by the Lucas number $x = L_{2n-1}$ and the Fibonacci number $y = F_{2n-1}$ for all $n \geq 1$.

(b) Show that the Pell equation $x^2 - 5y^2 = 4$ is solved by $x = L_{2n}$ and $y = F_{2n}$ for all $n \geq 1$.

[*Hint*: $x^2 - 5y^2 = \left(x + \sqrt{5}y\right)\left(x - \sqrt{5}y\right)$ *Comment*: It can be shown these are the *only* solutions of these Pell equations.].

Answer

(a)
$$\left(L_{2n-1}^2 - 5F_{2n-1}^2\right) = \left(L_{2n-1} + \sqrt{5}F_{2n-1}\right)\left(L_{2n-1} - \sqrt{5}F_{2n-1}\right)$$
$$= \left(\varphi^{2n-1} + \hat{\varphi}^{2n-1} + \varphi^{2n-1} - \hat{\varphi}^{2n-1}\right)\left(\varphi^{2n-1} + \hat{\varphi}^{2n-1} - \varphi^{2n-1} + \hat{\varphi}^{2n-1}\right)$$
$$= \left(2\varphi^{2n-1}\right)\left(2\hat{\varphi}^{2n-1}\right) = 4\left(\varphi\hat{\varphi}\right)^{2n-1} = 4(-1)^{2n-1} = -4$$

(b)
$$\left(L_{2n}^2 - 5F_{2n}^2\right) = \left(L_{2n} + \sqrt{5}F_{2n}\right)\left(L_{2n} - \sqrt{5}F_{2n}\right)$$
$$= \left(\varphi^{2n} + \hat{\varphi}^{2n} + \varphi^{2n} - \hat{\varphi}^{2n}\right)\left(\varphi^{2n} + \hat{\varphi}^{2n} - \varphi^{2n} + \hat{\varphi}^{2n}\right)$$
$$= \left(2\varphi^{2n}\right)\left(2\hat{\varphi}^{2n}\right) = 4\left(\varphi\hat{\varphi}\right)^{2n} = 4(-1)^{2n} = 4$$

5.7.15. The "four number game" (also called a Ducci sequence) is played this way: Choose a starting circular sequence of numbers (*a, b, c, d*) and at each step replace the sequence with the absolute values of the differences of adjacent entries: $(a, b, c, d) \to (|a-b|, |b-c|, |c-d|, |d-a|)$. For example, $(5, 350, 419, 37) \to (345, 69, 382, 32) \to (276, 313, 350, 313) \to (37, 37, 37, 37) \to (0, 0, 0, 0)$which shows that (0, 0, 0, 0) was reached in just four steps.

(a) Play several games, choosing starting values (*a, b, c, d*) that you believe may take more steps before reaching (0, 0, 0, 0).

(b) If *a, b, c,* and *d* are integers, prove that at least by the fourth step the 4-tuple has all even numbers. [*Hint*: Let *d* and *e* denote, respectively, odd and even numbers. For example, $(d, d, e, e) \to (e, d, e, d) \to (d, d, d, d) \to (e, e, e, e)$ shows all even numbers are reached in just three steps.]

(c) Use part (b) to explain why the game must end with (0, 0, 0, 0) in finitely many steps. [*Hint*: The number of steps to termination at (0, 0, 0, 0) is unchanged if (*a, b, c, d*) is replaced by (*a/m, b/m, c/m, d/m*), where *m* is a common factor of *a, b, c,* and *d*.]

(d) Play the four number game starting with any sequence of four consecutive numbers from the tribonacci sequence 0, 0, 1, 1, 2, 4, 7, 13, 24, 44, 81, 149, 274, 504, 927, 1705, 3136, 5768, 10609, 19513, 35890, 66012, … . Describe the pattern you should discover, and explain why the four number game can be made arbitrarily long.

(e) Let $\alpha \in (1, 2)$ be the real eigenvalue of the tribonacci recursion discussed in Example 5.20; that is, α satisfies the equation $\alpha^3 = \alpha^2 + \alpha + 1$. Explain why the four number game starting with $\left(1, \alpha, \alpha^2, \alpha^3\right)$ never terminates with all zeros.

Answer

(a) Examples will vary.

(b) Consider cases that depend on the number of odd entries in the starting 4-tuple. Clearly (*e, e, e, e*) is even in no steps at all, and $(d, d, d, d) \to (e, e, e, e)$ shows all even numbers appear after one step. When there are 3 odds and one even, we may as well assume that the even number appears first. All evens are then reached in 4 steps since $(e, d, d, d) \to (d, e, e, d) \to (d, e, d, e) \to (d, d, d, d) \to (e, e, e, e)$. This also shows that with a starting 4-tuple with two odds and two evens terminates in 3 steps if the odds are adjacent and terminates in 2 steps if the odds are between the evens. Finally, if there is one odd and three evens, all evens are reached in 4 steps since $(d, e, e, e) \to (d, e, e, d) \to (d, e, d, e) \to (d, d, d, d) \to (e, e, e, e)$.

(c) After the first step, the entries in the 4-tuple are all nonnegative, and at each additional step the numbers in the new 4-tuple are never larger than those in the previous 4-tuple. At least in every sequence of 4 more steps, all the numbers reached are even, so a factor of 2 can be divided out, and at least by 8 steps there is a factor of 4 in common. In sufficiently many additional steps, the 4 tuple has a common factor of any power of 2 whatsoever. Thus, we must reach the all 0 4-tuple in finitely many steps, since 0 is the only integer divisible by all powers of 2.

(d) For example $(81, 149, 274, 504) \to (68, 125, 230, 423) \to (57, 105, 193, 355) \to (48, 88, 162, 298) = 2\,(24, 44, 81, 149)$, so $(T_{10}, T_{11}, T_{12}, T_{13})$ becomes $2\left(T_8, T_9, T_{10}, T_{11}\right)$ in three steps.

In general, repeatedly using the recursion $T_{n+3} = T_{n+2} + T_{n+1} + T_n$

$$\begin{aligned}
(T_n, T_{n+1}, T_{n+2}, T_{n+3}) &\to \left(T_{n-1} + T_{n-2}, T_n + T_{n-1}, T_{n+1} + T_n, T_{n+1} + T_{n+2}\right) \\
&\to \left(T_n - T_{n-2}, T_{n+1} - T_{n-1}, T_{n+2} - T_n, T_{n+1} + T_{n+2} - T_{n-1} - T_{n-2}\right) \\
&= \left(T_{n-1} + T_{n-3}, T_n + T_{n-2}, T_{n+1} + T_{n-1}, 2T_{n+1} + T_n - T_{n-2}\right) \\
&\to \left(T_n + T_{n-2} - T_{n-1} - T_{n-3}, T_{n+1} + T_{n-1} - T_n - T_{n-2}, T_{n+1}\right. \\
&\quad \left. + T_n - T_{n-2} - T_{n-1}, 2T_{n+1} + T_n - T_{n-2} - T_{n-1} - T_{n-3}\right) \\
&= 2\left(T_{n-2}, T_{n-1}, T_n, T_{n+1}\right).
\end{aligned}$$

so $\left(T_n, T_{n+1}, T_{n+2}, T_{n+3}\right) \to\to\to 2\left(T_{n-2}, T_{n-1}, T_n, T_{n+1}\right)$. We see that the number of steps can be made arbitrarily large by letting n be sufficiently large.

(e) Since $\left(1, \alpha, \alpha^2, \alpha^3\right) \to \left(\alpha - 1, \alpha^2 - \alpha, a^3 - \alpha^2, \alpha^3 - 1\right) = (\alpha - 1)\left(1, \alpha, \alpha^2, \alpha^2 + \alpha + 1\right) = (\alpha - 1)\left(1, \alpha, \alpha^2, \alpha^3\right)$each step gives a proportional 4-tuple, and after n steps the 4-tuple reached is $(\alpha - 1)^n\left(1, \alpha, \alpha^2, \alpha^3\right)$, which is never all zeroes.

6

SPECIAL NUMBERS

PROBLEM SET 6.2

6.2.1. Let $p(n) = 3n^2 - n + 2$.

(a) Calculate the difference table.

(b) Express $p(n)$ as a linear sum of binomial coefficients.

(c) Express $\sum_{m=0}^{n} p(m)$ as a linear sum of binomial coefficients.

Answer

(a)

$$\begin{array}{ccccccc} 2 & & 4 & & 12 & & 26 \\ & 2 & & 8 & & 14 & \\ & & 6 & & 6 & & \\ & & & 0 & & & \end{array}$$

(b) $p(n) = 2\binom{n}{0} + 2\binom{n}{1} + 6\binom{n}{2}$

Solutions Manual to Accompany Combinatorial Reasoning: An Introduction to the Art of Counting, First Edition. Duane DeTemple and William Webb.

(c) By Corollary 6.4,

$$\sum_{m=0}^{n} p(m) = 2\binom{n+1}{1} + 2\binom{n+1}{2} + 6\binom{n+1}{3}$$

6.2.3. The sequence of centered triangular numbers is 1, 4, 10, 19, 31, … . Determine the quadratic polynomial that gives this sequence.

Answer
The polynomial can be determined from the difference table.

$$\begin{array}{ccccccccc}
1 & & 4 & & 10 & & 19 & & 31 \\
 & 3 & & 6 & & 9 & & 12 & \\
 & & 3 & & 3 & & 3 & & \\
 & & & 0 & & 0 & & &
\end{array}$$

$$p(n) = \binom{n}{0} + 3\binom{n}{1} + 3\binom{n}{2} = 1 + 3n + 3\frac{n(n-1)}{2} = \frac{3n^2 + 3n + 2}{2}$$

6.2.5. The diagram below shows the interior regions formed by the diagonals of convex n-gons for $n = 1, 2, 3, 4, 5$, and 6. Find a quartic polynomial $p(n)$ that gives the maximum number of regions formed by the diagonals of a convex n-gon.

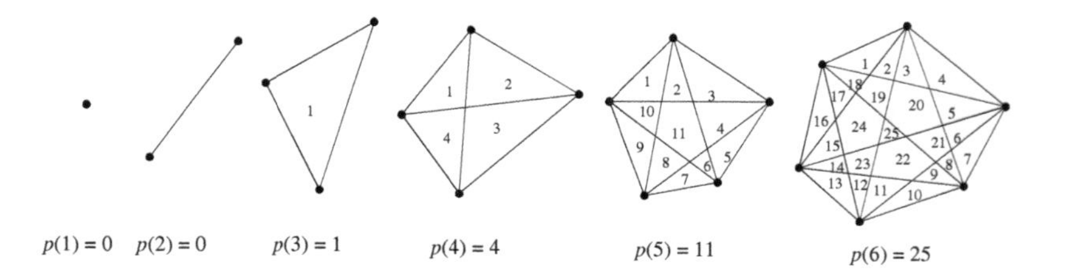

Answer
Form the difference table, where $p(0) = 1$ by extending the table to the left.

$$\begin{array}{ccccccccccccc}
1 & & 0 & & 0 & & 1 & & 4 & & 11 & & 25 \\
 & -1 & & 0 & & 1 & & 3 & & 7 & & 14 & \\
 & & 1 & & 1 & & 2 & & 4 & & 7 & & \\
 & & & 0 & & 1 & & 2 & & 3 & & & \\
 & & & & 1 & & 1 & & 1 & & & & \\
 & & & & & 0 & & 0 & & & & &
\end{array}$$

By Theorem 6.3, $p(n) = \binom{n}{0} - \binom{n}{1} + \binom{n}{2} + \binom{n}{4}$. Using Pascal's identity, this can be written in the form $p(n) = 1 - n + \binom{n}{2} + \binom{n}{4} = -\binom{n-1}{1} + \binom{n-1}{1} + \binom{n-1}{2} + \binom{n-1}{3} + \binom{n-1}{4} = \binom{n-1}{2} + \binom{n-1}{3} + \binom{n-1}{4}$ which is the sum of the third, fourth, and fifth entries in row $n-1$ of Pascal's triangle. For example, row 5 of Pascal's triangle is 1 5 10 10 5 1 so $p(6) = 10 + 10 + 5 = 25$.

6.2.7. Euler's polynomial in Problem 6.2.6 often gives a prime number. However, prove there is no nonconstant polynomial $P(n)$ whose values at $n = 1, 2, 3, \ldots$ are all prime numbers. [*Hint*: Let $P(1) = p$, where p is a prime number. Explain why $P(1 + kp)$ is divisible by p for any positive integer k.]

Answer

Let $P(x) = \sum_{j=0}^{m} b_j x^j$ so $P(1) = \sum_{j=0}^{m} b_j = p$ for some prime number p. Then

$$\begin{aligned} P(1+kp) &= \sum_{j=0}^{m} b_j (1+kp)^j \\ &= \sum_{j=0}^{m} b_j + (\text{terms each containing a positive power of } p) \\ &= P(1) + (\text{multiple of } p) = p + (\text{multiple of } p) = (\text{multiple of } p) \end{aligned}$$

Since $P(1 + kp)$ is a prime number that is a multiple of p, then $P(1 + kp) = p$ for infinitely many k. But this forces P to be the constant polynomial $P(n) = p$ for all n.

6.2.9. If $p(n) = \sum_{j=0}^{k} c_j (n)_j$ is 0 for all integers $n \geq 0$, prove that all of the coefficients c_j are 0.

Answer

If not all of the coefficients c_j are 0, we may as well assume that k is the largest index for which $c_k \neq 0$. The polynomial then has the form $p(n) = c_k n^k + \hat{p}(n)$ where $\hat{p}(n)$ is a polynomial of degree at most $k - 1$. Since $p(n)$ is the 0 polynomial, all of its coefficients are 0, so $c_k = 0$, a contradiction. Therefore all of the coefficients c_j are 0. (see Problem 3.2.8(a).)

6.2.11. Recall that $c_n^{(r)} = c^{(r)}(n)$ denotes the nth centered r-gonal number, where $c^{(r)}(0) = 1$. For example, the centered pentagonal numbers $r = 5$ are shown for $n = 0, 1, 2, 3, 4$, and 5.

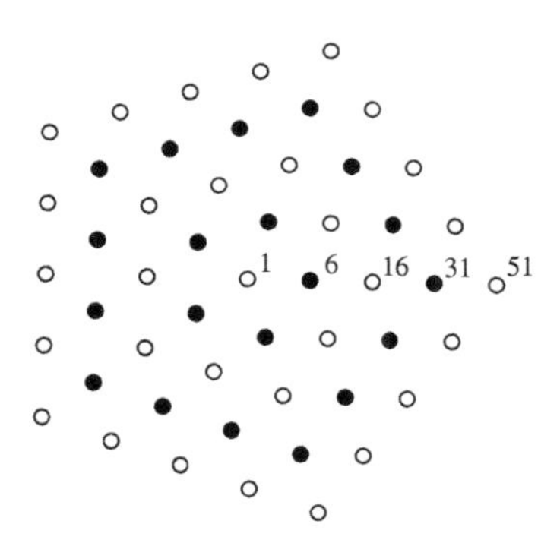

(a) Explain why $\Delta c^{(r)}(n) = (n+1)r$.

(b) Use part (a) to prove that $c^{(r)}(n) = 1 + r\binom{n+1}{2} = 1 + \dfrac{rn(n+1)}{2}$

Answer

(a) As seen from the diagram, $c^{(r)}(n+1)$ has $(n+1)r$ dots added to those of $c^{(r)}(n)$.

(b) Since $\Delta c^{(r)}(n) = (n+1)r$ then $\Delta^2 c^{(r)}(n) = r$ and $\Delta^3 c^{(r)}(n) = 0$. Therefore $c^{(r)}(n)\,|_{n=0} = 1$, $\Delta c^{(r)}(n)\,|_{n=0} = r$, $\Delta^2 c^{(r)}(n)\,|_{n=0} = r$ and $\Delta^3 c^{(r)}(n)\,|_{n=0} = 0$ so it follows from Theorem 6.3 that $c^{(r)}(n) = 1 + r\binom{n}{1} + r\binom{n}{2} = 1 + r\binom{n+1}{2}$.

6.2.13. Use the principle of inclusion-exclusion (PIE) to prove that

$$T(k,j) = j!\left\{ {k \atop j} \right\} = \sum_{t=0}^{j} (-1)^t \binom{j}{t} (j-t)^k$$

and show this is equivalent to the identity

$$\left\{ {k \atop j} \right\} = \frac{1}{j!} \sum_{i=0}^{j} (-1)^{i+j} \binom{j}{i} i^k$$

given by equation (6.9).

Answer

By definition, $T(k,j)$ is the number of ways to assign k people to j distinct rooms, where no room is left unoccupied. But $j!\left\{ {k \atop j} \right\}$ gives the number of assignments as well, since there are $\left\{ {k \atop j} \right\}$ ways to form j nonempty groups of the k people and $j!$ ways to place these groups into j distinct rooms. Let U be the set of all room assignments, possibly leaving some rooms empty. Then $|U| = j^k$, since each of the k people is given j choices of a room. Let $A_r \subseteq U$ be the subset of room assignments that leave room r empty and possibly others as well, so $|A_r| = (j-1)^k$. Similarly, $|A_r \cap A_s| = (j-2)^k$ is the number of assignments that leave at least the two rooms r and s empty, where $r \neq s$. There are $\binom{j}{2}$ ways to choose the two rooms that are left unoccupied. In general, there are $\binom{j}{t}$ ways to choose a subset of t rooms that must be left empty, and $(j-t)^k$ ways to assign k people to the remaining rooms. The number of room assignments that leave no room empty is, by PIE, then

$$\left|\bar{A}_1 \cap \bar{A}_2 \cap \cdots \cap \bar{A}_j\right| = j^k - j(j-1)^k + \binom{j}{2}(j-2)^k - \cdots$$

$$= \sum_{t=0}^{j} (-1)^t \binom{j}{t} (j-t)^k .$$

The alternate form is obtained by replacing the summation variable with $i = j - t$.

6.2.15. Prove that

$$\sum_{i=0}^{k} (-1)^i \left\{ {k \atop i} \right\} \left[{i \atop j} \right] = (-1)^k \delta_{j,k}$$

Answer

Using equations (6.10) and (6.14),

$$x^k = \sum_{i=0}^{k} \left\{ {k \atop i} \right\} (x)_i = \sum_{i=0}^{k} \left\{ {k \atop i} \right\} \sum_{j=0}^{i} (-1)^{i-j} \left[{i \atop j} \right] x^j$$

$$= \sum_{j=0}^{k} (-1)^j \left(\sum_{i=0}^{k} (-1)^i \left\{ {k \atop i} \right\} \left[{i \atop j} \right] \right) x^j$$

Comparing coefficients of x^j gives the result.

6.2.17. Prove the hockey stick-like identity for Stirling numbers of the second kind:

$$\left\{ {j+k+1 \atop j} \right\} = \sum_{i=0}^{j} i \left\{ {k+i \atop i} \right\}, \; j,k \geq 0.$$

Answer

The formula is easily seen to be true for $j = 0$ and $j = 1$ for every $k \geq 0$. Now suppose the formula holds for some $j \geq 1$ and all $k \geq 0$. Then, using the triangle identity for the Stirling numbers of the second kind, we obtain

$$\begin{aligned}
\left\{ {j+1+k+1 \atop j+1} \right\} &= (j+1)\left\{ {j+k+1 \atop j+1} \right\} + \left\{ {j+k+1 \atop j} \right\} \\
&= (j+1)\left\{ {j+k+1 \atop j+1} \right\} + \sum_{i=0}^{j} i \left\{ {k+i \atop i} \right\} \\
&= \sum_{i=0}^{j+1} i \left\{ {k+i \atop i} \right\}
\end{aligned}$$

so the identity is true by mathematical induction.

6.2.19. Prove the hockey stick-like identity for Stirling numbers of the first kind.

$$\left[{j+k+1 \atop j} \right] = \sum_{i=0}^{j} (k+i) \left[{k+i \atop i} \right]$$

Answer

The formula is easily seen to be true for $j = 0$ and $j = 1$ for every $k \geq 0$. Now suppose the formula holds for some $j \geq 1$ and all $k \geq 0$. Then, using the triangle identity for the Stirling numbers of the first kind,

$$\begin{aligned}
\left[{j+1+k+1 \atop j+1} \right] &= (j+k+1)\left[{j+k+1 \atop j+1} \right] + \left[{j+k+1 \atop j} \right] \\
&= (j+k+1)\left[{j+k+1 \atop j+1} \right] + \sum_{i=0}^{j} (k+i) \left[{k+i \atop i} \right] \\
&= \sum_{i=0}^{j+1} (k+i) \left[{k+i \atop i} \right]
\end{aligned}$$

so the identity is true by mathematical induction.

6.2.21. Give a combinatorial proof that

$$\left\{ {k+1 \atop j+1} \right\} = \sum_{i=j}^{k} \binom{k}{i} \left\{ {i \atop j} \right\}$$

Answer

There are $\left\{ {k+1 \atop j+1} \right\}$ ways that $k + 1$ distinct toys can be placed into $j + 1$ identical toy boxes with no box left empty. Now suppose i toys are not in the same box as toy $k + 1$. These i toys can be chosen and placed in j toy boxes in $\binom{k}{i} \left\{ {i \atop j} \right\}$ ways, with the remaining $k - i$ toys placed together with toy $k + 1$ in the other toy box. Summing over all i gives all of the placements of the $k + 1$ distinct toys into $j + 1$ identical toy boxes with no box empty.

6.2.23. Give a combinatorial proof that

$$\binom{r+b}{r} \left\{ {k \atop r+b} \right\} = \sum_{j=0}^{k} \binom{k}{j} \left\{ {j \atop r} \right\} \left\{ {k-j \atop b} \right\}$$

[*Hint*: Ask a question about red and blue toy boxes that can be answered in two ways.]

Answer

"In how many ways can k distinct toys be placed into r red and b blue toy boxes, where toy boxes of the same color are identical and no box is left empty?"

Answer 1. The k toys can be put into $r + b$ identical unpainted toy boxes in $\left\{ {k \atop r+b} \right\}$ ways and then r of the $r + b$ boxes can be chosen to be painted red in $\binom{r+b}{r}$ ways, with the remaining toy boxes painted blue. One answer is therefore $\left\{ {k \atop r+b} \right\} \binom{r+b}{r}$.

Answer 2. First paint r toy boxes red and the rest blue. Next suppose that, in $\binom{k}{j}$ ways, j toys are chosen for the r red toys boxes and placed in them in $\left\{ {j \atop r} \right\}$ ways. This leaves $\left\{ {k-j \atop b} \right\}$ ways to place the remaining toys in the blue toy boxes. That is, there are $\binom{k}{j} \left\{ {j \atop r} \right\} \left\{ {k-j \atop b} \right\}$ placements with j toys in the red boxes. Summing over j gives all of the placements.

Equating the two answers gives the identity.

6.2.25. Give a combinatorial proof using the DIE method of Section 4.2 to prove that

$$\sum_{j=1}^{k} (-1)^j \begin{bmatrix} k \\ j \end{bmatrix} = 0, \; k \geq 2$$

Answer

D. The unsigned sum $\sum_{j=1}^{k} \begin{bmatrix} k \\ j \end{bmatrix}$ counts all ways to seat k people around k identical circular tables, where j is the number of occupied tables in the seating.

I. If persons 1 and $k \geq 2$ are at different tables with the respective seatings, $(1, a_1, a_2, \ldots, a_r)$ and $(k, b_1, b_2, \ldots, b_s)$, combine the people at two tables to give the seating $(1, a_1, a_2, \ldots, a_r, k, b_1, b_2, \ldots, b_s)$ that uses one fewer table. In the opposite direction, if persons 1 and $k \geq 2$ are at the same table with the seating, $(1, a_1, a_2, \ldots, a_r, k, b_1, b_2, \ldots, b_s)$, these people can be reseated using one additional table with the seatings $(1, a_1, a_2, \ldots, a_r)$ and $(k, b_1, b_2, \ldots, b_s)$.

E. An exception would occur only if persons 1 and k are seated at the same table and there are no tables left unused. But there are $k \geq 2$ tables, so this situation cannot occur and therefore there are no exceptions.

There are no exceptional cases, so the sum is 0.

PROBLEM SET 6.3

6.3.1. Suppose that the harmonic series has a finite sum H, in which case the sums of the odd terms $D = 1 + \frac{1}{3} + \frac{1}{5} + \cdots$ and even terms $E = \frac{1}{2} + \frac{1}{4} + \frac{1}{6} + \cdots$ would also be finite with $H = D + E$. Show that $2E = H$, and explain why there is a contradiction to the assumption that H is finite.

Answer

$2E = 2\left(\frac{1}{2} + \frac{1}{4} + \frac{1}{6} + \cdots\right) = 1 + \frac{1}{2} + \frac{1}{3} + \cdots = H$ so $D = H - E = E$. But $D > E$ since each term of D is strictly larger than the corresponding term of E. This is a contradiction, so the assumption that H is finite is false.

6.3.3. Prove that $H_N - H_n = \frac{1}{n+1} + \frac{1}{n+2} + \cdots + \frac{1}{N}$ is never an integer for $0 < n < N$.

Answer

The proof is the same as that for Theorem 6.27, once it is shown there is a unique integer k, $n + 1 \leq k \leq N$ whose degree of evenness is a maximum. Suppose that both $k = 2^j m_k$ and $k' = 2^j m'_k$ have the same maximum degree of evenness j, where $m_k < m'_k$. Since m_k and m'_k are both odd, there is an even integer $2m$ between them. But this means $2^j \cdot 2m$ is an integer between $n + 1$ and N that has a degree of evenness larger than j. This contradicts the assumption that j was the largest degree of evenness of all of the integers between $n + 1$ and N.

6.3.5. Use the generating function $\sum_{n=0}^{\infty} H_n x^n = -\frac{\ln(1-x)}{1-x}$ derived Problem 6.3.4 to prove that $\sum_{n=0}^{s-1} H_n = sH_s - s$ for $s \geq 1$.

Answer

Take the derivative of the generating function to get $\sum_{n=1}^{\infty} nH_n x^{n-1} = \frac{1}{(1-x)^2} - \frac{\ln(1-x)}{(1-x)^2} = \sum_{n=1}^{\infty} nx^{n-1} + \sum_{n=0}^{\infty}\left(\sum_{k=0}^{n} H_k\right)x^n$. Equating coefficients of x^{s-1} gives the formula.

6.3.7. The Euler constant was defined by $\gamma = \lim_{n\to\infty}\left(H_n - \ln(n+1)\right)$. Prove that

(a) $\gamma = \lim_{n\to\infty}\left(H_n - \ln n\right)$

(b) $\gamma = \lim_{n\to\infty}\left(H_n - \ln\left(n + \frac{1}{2}\right)\right)$

Answer

(a)
$$\begin{aligned}\lim_{n\to\infty}\left(H_n - \ln n\right) &= \lim_{n\to\infty}\left(H_n - \ln(n+1) + \ln(n+1) - \ln n\right)\\ &= \lim_{n\to\infty}\left(H_n - \ln(n+1)\right) + \lim_{n\to\infty}\ln\left(\frac{n+1}{n}\right)\\ &= \gamma + \ln 1 = \gamma\end{aligned}$$

(b) $H_n - \ln(n+1) < H_n - \ln\left(n + \frac{1}{2}\right) < H_n - \ln n$ so taking the limit shows that $\gamma \leq \lim_{n\to\infty}\left(H_n - \ln\left(n + \frac{1}{2}\right)\right) \leq \gamma$.

6.3.9. Show that

$$\sum_{n=1}^{k} \frac{2n+1}{n(n+1)} = 2H_k - \frac{k}{k+1}$$

for all $k \geq 1$.

Answer

$$\begin{aligned}\sum_{n=1}^{k} \frac{2n+1}{n(n+1)} &= \sum_{n=1}^{k}\left(\frac{1}{n} + \frac{1}{n+1}\right) = H_k + H_{k+1} - 1 = 2H_k + \frac{1}{k+1} - 1\\ &= 2H_k - \frac{k}{k+1}\end{aligned}$$

6.3.11. Use summation by parts to prove these summation formulas.

(a) $\sum_{n=0}^{k-1} \frac{1}{(n+1)(n+2)} = \frac{k}{k+1}$

(b) $\sum_{n=0}^{k-1} \frac{H_n}{(n+1)(n+2)} = \frac{k-H_k}{k+1}$ [*Hint*: part (a) can be helpful]

Answer

(a) $\sum_{n=0}^{k-1} \frac{1}{(n+1)(n+2)} = \sum_{n=0}^{k-1} \left(\frac{1}{n+1} - \frac{1}{n+2}\right) = -\sum_{n=0}^{k-1} \Delta\left(\frac{1}{n+1}\right) = -\frac{1}{n+1}\Big|_{n=0}^{n=k} = 1 - \frac{1}{k+1} = \frac{k}{k+1}$

(b) $\sum_{n=0}^{k-1} \frac{H_n}{(n+1)(n+2)} = -\sum_{n=0}^{k-1} H_n\left(\frac{1}{n+2} - \frac{1}{n+1}\right)$

$= -\sum_{n=0}^{k-1} H_n \Delta\left(\frac{1}{n+1}\right) = -\frac{H_n}{n+1}\Big|_{n=0}^{n=k} + \sum_{n=0}^{k-1} \Delta(H_n) E\left(\frac{1}{n+1}\right)$

$= -\frac{H_k}{k+1} + \sum_{n=0}^{k-1}\left(\frac{1}{n+1}\right)\left(\frac{1}{n+2}\right) = -\frac{H_k}{k+1} + \frac{k}{k+1}$ [by part (a)]

6.3.13. Use summation by parts to prove that

$$\sum_{n=0}^{k-1} \binom{n}{r} H_n = \binom{k}{r+1}\left(H_k - \frac{1}{r+1}\right)$$

[*Hint*: Use $\binom{n}{r} = \Delta\binom{n}{r+1}$, the *committee selection with chair* identity $(r+1)\binom{n+1}{r+1} = (n+1)\binom{n}{r}$, and the hockey stick identity.]

Answer

$$\sum_{n=0}^{k-1} \binom{n}{r} H_n = \sum_{n=0}^{k-1} \Delta\binom{n}{r+1} H_n = \binom{n}{r+1} H_n \Big|_{n=0}^{n=k} - \sum_{n=0}^{k-1} \Delta H_n E\binom{n}{r+1}$$

$$= \binom{k}{r+1} H_k - \sum_{n=0}^{k-1} \frac{1}{n+1}\binom{n+1}{r+1}$$

$$= \binom{k}{r+1} H_k - \frac{1}{r+1}\sum_{n=0}^{k-1}\binom{n}{r} \text{ [by the committee with chair identity]}$$

$$= \binom{k}{r+1} H_k - \frac{1}{r+1}\binom{k}{r+1} \text{ [by the hockey stick identity]}$$

6.3.15. Jack's and Jill's tastes in jellybeans are exactly like Ronnie's described in Example 6.28—only green ones are eaten and red ones are returned to the bag. Jack has a bag with 10 red and 1 green jellybean, and Jill's bag has 10 green and 1 red jellybean. Who can be expected to eat all of the green jellybeans in their respective bags first?

Answer

$$\begin{aligned} E_{Jack}(1,10) &= 10 + H_{10} \approx 10 + \gamma + \ln 10.5 = 10 + 0.577\ldots + 2.351 \\ &= 12.928\ldots \\ E_{Jill}(10,1) &= 1 + 10H_1 = 11 \end{aligned}$$

Jill can be expected to finish eating all 10 of her green jellybeans a couple of days before Jack finally eats his one green jellybean.

PROBLEM SET 6.4

6.4.1. Use Definition 6.35 to derive the Bernoulli polynomial $B_5(t)$, assuming that $B_4(t) = t^4 - 2t^3 + t^2 - \frac{1}{30}$ has already been calculated.

Answer

$B_5(t) = 5\int\left(t^4 - 2t^3 + t^2 - \frac{1}{30}\right)dt = t^5 - \frac{5}{2}t^4 + \frac{5}{3}t^3 - \frac{1}{6}t + C$ where the constant of integration C is determined by the equation

$$0 = \int_0^1 B_5(t)\,dt = \int_0^1 \left(t^5 - \frac{5}{2}t^4 + \frac{5}{3}t^3 - \frac{1}{6}t + C\right)dt = \frac{1}{6}$$

$$-\frac{1}{2} + \frac{5}{12} - \frac{1}{12} + C = 0 + C$$

That is, $C = 0 = B_5$.

6.4.3. Prove the difference formula $B_n(t+1) - B_n(t) = nt^{n-1}$.

[*Hint*: Define the polynomials $p_0(t) = 0$ and $p_n(t) = B_n(t+1) - B_n(t) - nt^{n-1}$ for $n \geq 1$ and take their derivatives.]

Answer

Since both $B_n(t+1)$ and $B_n(t)$ are monic polynomials of degree n, $p_n(t)$ is a polynomial of degree at most $n-1$ with

$$p_0(t) = 0,\ p_1(0) = B_1(1) - B_1(0) - 1 = \frac{1}{2} - \left(-\frac{1}{2}\right) - 1 = 0,$$
$$\text{and } p_n(0) = B_n(1) - B_n(0) - n0^{n-1} = 0 \text{ for all } n \geq 2.$$

Therefore

$$\begin{aligned} p_n'(t) = B_n'(t+1) - B_n'(t) - n(n-1)t^{n-2} &= n\left(B_{n-1}(t+1)\right. \\ \left. -B_{n-1}(t) - (n-1)t^{n-2}\right) &= np_{n-1}(t) \end{aligned}$$

which shows by induction that $p_n^{(k)}(0) = 0$ for all $k \geq 0$. We conclude that $p_n(t) = 0$.

6.4.5. Prove the addition formula $B_n(t+h) = \sum_{k=0}^{n} \binom{n}{k} B_k(t) h^{n-k}$, $n \geq 0$.

[*Hint*: Use the generating function for the Bernoulli polynomials given by Theorem 6.37.]

Answer

$$B_n(t+h) = \left[\frac{x^n}{n!}\right] \frac{xe^{x(t+h)}}{e^x - 1} = \left[\frac{x^n}{n!}\right] \left(\frac{xe^{xt}}{e^x - 1} e^{xh}\right)$$

$$= \left[\frac{x^n}{n!}\right] \left(\frac{xe^{xt}}{e^x - 1}\right) \left(e^{xh}\right) = \left[\frac{x^n}{n!}\right] \left(\sum_{r=0}^{\infty} B_r(t) \frac{x^r}{r!}\right) \left(\sum_{s=0}^{\infty} h^s \frac{x^s}{s!}\right)$$

$$= \left[\frac{x^n}{n!}\right] \sum_{n=0}^{\infty} \left(\sum_{k=0}^{n} \binom{n}{k} B_k(t) h^{n-k}\right) \frac{x^n}{n!} = \sum_{k=0}^{n} \binom{n}{k} B_k(t) h^{n-k}$$

6.4.7. Use the multiplication formula of Problem 6.4.6 to prove that

$$B_{2n}\left(\frac{1}{3}\right) = B_{2n}\left(\frac{2}{3}\right) = \left(3^{2n-1} - 1\right) B_{2n}$$

Answer

Recall that $B_{2n}\left(\frac{1}{3}\right) = B_{2n}\left(\frac{2}{3}\right)$ by the even symmetry of $B_{2n}(t)$ about $t = 1/2$. The multiplication formula, with $m = 3$ and $t = 0$, then gives $B_{2n} = 3^{2n-1}\left(B_{2n} + B_{2n}\left(\frac{1}{3}\right) + B_{2n}\left(\frac{2}{3}\right)\right)$ so $B_{2n}\left(\frac{1}{3}\right) = \frac{1}{2}\left(3^{1-2n} - 1\right) B_{2n}$.

6.4.9. Use equation (6.48) to derive the formula for $\sigma_5(n)$, assuming that $\sigma_1(n), \sigma_2(n), \sigma_3(n), \sigma_4(n)$ are known.

Answer

$$\sigma_5(n) = \frac{1}{5+1}\left(n^{5+1} - \sum_{j=0}^{5-1} \binom{5+1}{j} \sigma_j(n)\right)$$

$$= \frac{1}{6}\left(n^6 - \sigma_0(n) - 6\sigma_1(n) - 15\sigma_2(n) - 20\sigma_3(n) - 15\sigma_4(n)\right)$$

$$= \frac{1}{6}n^6 - \frac{1}{6}(n) - \frac{6}{6}\left(\frac{1}{2}n^2 - \frac{1}{2}n\right) - \frac{15}{6}\left(\frac{1}{3}n^3 - \frac{1}{2}n^2 + \frac{1}{6}n\right)$$

$$-\frac{20}{6}\left(\frac{1}{4}n^4 - \frac{1}{2}n^3 + \frac{1}{4}n^2\right) - \frac{15}{6}\left(\frac{1}{5}n^5 - \frac{1}{2}n^4 + \frac{1}{3}n^3 - \frac{1}{30}n\right)$$

$$= \frac{1}{6}n^6 - \left(\frac{2-6+5-1}{12}\right)n - \left(\frac{6-15+10}{12}\right)n^2$$
$$- \left(\frac{5-10+5}{6}\right)n^3 - \left(\frac{10-15}{12}\right)n^4 - \frac{1}{2}n^5$$
$$= \frac{1}{6}n^6 - (0)\,n - \left(\frac{1}{12}\right)n^2 - (0)\,n^3 + \left(\frac{5}{12}\right)n^4 - \frac{1}{2}n^5$$

6.4.11. Prove that

(a) $\cot x + \tan\dfrac{x}{2} = \dfrac{1}{\sin x}$

[*Hint*: $\cot 2y = \frac{1}{2}(\cot y - \tan y)$ by Problem 6.4.10(a)]

(b) $\dfrac{x}{\sin x} = \displaystyle\sum_{n=0}^{\infty}(-1)^{n-1}\left(4^n - 2\right)B_{2n}\frac{x^{2n}}{(2n)!}$

Answer

(a) $\cot 2y + \tan y = \frac{1}{2}(\cot y - \tan y) + \tan y = \frac{1}{2}(\cot y + \tan y) = \dfrac{\cos^2 y + \sin^2 y}{2\sin y\cos y} = \dfrac{1}{\sin 2y}$ so letting $2y = x$ gives the identity needed.

(b) By part (a) and Problem 6.4.10(b),

$$\frac{x}{\sin x} = x\cot x + x\tan\frac{x}{2}$$
$$= \sum_{n=0}^{\infty}(-1)^n 4^n B_{2n}\frac{x^{2n}}{(2n)!} + \sum_{n=0}^{\infty}(-1)^{n-1}4^n\left(4^n - 1\right)B_{2n}\frac{x^{2n}}{(2n)!}$$
$$= \sum_{n=0}^{\infty}(-1)^{n-1}4^n\left(4^n - 2\right)B_{2n}\frac{x^{2n}}{(2n)!}.$$

PROBLEM SET 6.5

6.5.1. Give the spreadsheet formula to be entered in cell C3 that calculates the Eulerian number $\left\langle {2 \atop 1} \right\rangle$ and can then be copied into the remaining cells to generate the Eulerian numbers shown in Table 6.6.

◢	A	B	C	
1	$k\backslash i$	0	1	
2	1	1	–	
3	2	1	1	
4	3	1	4	
5	4	1	11	

Answer

$$C3 = (C\$1 + 1) * C2 + (\$A3 - C\$1) * B2$$

6.5.3. Give a combinatorial proof of $\left\langle {k \atop 1} \right\rangle = 2^k - k - 1$ using the fact that $\left\langle {k \atop i} \right\rangle$ is the number of permutations of $[k]$ with exactly i ascents. [*Hint.* Let A be the nonempty set of descending values preceding the one ascent, and B the set of descending values following the ascent. Now count the number of sets A and B.]

Answer

(a) Choose a nonempty subset $A \subseteq [k]$ and let B be its complement. If $\pi = a_1 a_2 \cdots a_{k-j} b_j \cdots b_2 b_1$ where $a_i \in A$, $a_1 > a_2 > \cdots > a_{k-j}$ and $b_i \in B$, $b_j > \cdots > b_2 > b_1$, then π is a permutation of $[k]$ with exactly one ascent except in the k cases for which $A = \{k, k-1, \ldots, j+1\}$ for $j = 0, 1, \ldots, k-1$. Since there are $2^k - 1$ nonempty subsets A, there are $2^k - 1 - k$ permutations of $[k]$ with exactly one ascent.

6.5.5. Show that the generating function for the Eulerian numbers $\left\langle {k \atop 2} \right\rangle$ in column $i = 2$ of Euler's triangle in Table 6.6 is

$$\sum_{k=1}^{\infty} \left\langle {k \atop 2} \right\rangle x^k = \frac{x^3 \left(1 + x - 4x^2\right)}{(1-3x)(1-2x)^2(1-x)^3}$$

$\left[\textit{Hint:} \text{By equation (6.58)}, \left\langle {k \atop 2} \right\rangle = 3^k - (k+1)\,2^k - \binom{k+1}{2} \right]$

Answer

Setting $\left\langle {0 \atop 2} \right\rangle = \left\langle {1 \atop 2} \right\rangle = 0,$

$$\sum_{k=0}^{\infty} \left\langle {k \atop 2} \right\rangle x^k = \sum_{k=0}^{\infty} (3x)^k - \sum_{k=0}^{\infty} (k+1)(2x)^k - \sum_{k=0}^{\infty} \binom{k+1}{2} x^k$$

$$= \frac{1}{1-3x} - \frac{1}{(1-2x)^2} - \frac{x}{(1-x)^3}$$

which simplifies to the desired OGF when the terms are placed over a common denominator.

6.5.7. **(a)** Give a combinatorial proof of the binomial coefficient identity

$$(n+1)\binom{m+1}{n} = \binom{m}{n} + (m+2)\binom{m}{n-1}.$$

(b) Define $\hat{C}(m,n) = \binom{-m-2}{n-1}$. Show that $\hat{C}(m,n) = n\hat{C}(m-1,n) + m\hat{C}(m,n-1)$.

(This is the same triangle identity satisfied by the coefficients $E(m,n) = \left\langle {m+n-1 \atop n-1} \right\rangle$ introduced in Problem 6.5.6.)

Answer

(a) Use a committee selection model and ask: "In how many ways can a committee of size n, either with or without a chair, be chosen from a club of m men and one woman?"

Answer 1. There are $\binom{m+1}{n}$ ways to select the committee, and $n+1$ ways either to choose one committee member as chair or else opt not to have a chair. Thus, there are $(n+1)\binom{m+1}{n}$ committees, the left side of the proposed identity.

Answer 2. There are $\binom{m}{n}$ committees of n men and with no chair; $m\binom{m}{n-1}$ committees chaired by a man with the remaining $n-1$ committee members chosen from the other m club members; $2\binom{m}{n-1}$ committees that include $n-1$ men, none a chair, together with the woman either as chair or not. This gives a total of $\binom{m}{n} + (m+2)\binom{m}{n-1}$ committees.

Equating the two answers proves the identity.

(b) The polynomial identity from part (a) holds for all integers m, so

$$\hat{C}(m,n) = \binom{-m-2}{n-1} = (n-1+1)\binom{-m-1}{n-1} - (-m-2+2)\binom{-m-2}{n-2}$$
$$= n\hat{C}(m-1,n) + m\hat{C}(m,n-1)$$

6.5.9. A *run* in a permutation of $[k]$ is a maximal increasing list of one or more consecutive numbers of the permutation. For example, the runs in the six permutations of [3] are shown with a vertical bar following the last number of a run:

$$(1,2,3|),(2,3|,1|),(3|,1,2|),(1,3|,2|),(2|,1,3|),(3|,2|,1|)$$

Prove that there are $\left\langle {k \atop i-1} \right\rangle$ permutations of $[k]$ with exactly i runs.

Answer

Consider all of the permutations with i runs. Each of these permutations has one run at the end of the permutation, and the other $i-1$ runs occur at a descent of the permutation. That is, $\ldots a|b, \ldots,$ where $a > b$. This means the permutation has $i-1$ descents, and therefore $k-1-(i-1)=k-i$ ascents. This accounts for all of the $\left\langle {k \atop k-i} \right\rangle$ permutations of $[k]$ with $k-i$ ascents. But using the symmetry relation (6.63), we see that the number of permutations with i runs is $\left\langle {k \atop k-i} \right\rangle = \left\langle {k \atop k-1-(k-i)} \right\rangle = \left\langle {k \atop i-1} \right\rangle$.

6.5.11. **(a)** Verify that the Euler polynomials $P_k(x) = (1-x)^{k+1} \sum_{n=1}^{\infty} n^k x^n$ satisfy the recursion relation $P_{k+1}(x) = x(1-x)P'_k(x) + (k+1)xP_k(x)$, $k \geq 1$.

(b) Use part (a) to obtain $P_2(x)$, $P_3(x)$ and $P_4(x)$ starting from $P_1(x) = x$.

Answer

(a) Differentiate $P_k(x) = (1-x)^{k+1} \sum_{n=1}^{\infty} n^k x^n$ and multiply the result by $x(1-x)$:

$$\begin{aligned} x(1-x)P'_k(x) &= (1-x)^{k+2} \sum_{n=1}^{\infty} n^{k+1}x^n - (k+1)x(1-x)^{k+1} \sum_{n=1}^{\infty} n^k x^n \\ &= P_{k+1}(x) - (k+1)xP_k(x) \end{aligned}$$

(b)

$$\begin{aligned} P_2(x) &= x(1-x)(1) + 2x(x) = x - x^2 + 2x^2 = x + x^2 \\ P_3(x) &= x(1-x)(1+2x) + 3x\left(x+x^2\right) \\ &= x + x^2 - 2x^3 + 3x^2 + 3x^3 = x + 4x^2 + x^3 \\ P_4(x) &= x(1-x)\left(1+8x+3x^2\right) + 4x\left(x+4x^2+x^3\right) \\ &= x + 7x^2 - 5x^3 - 3x^4 + 4x^2 + 16x^3 + 4x^4 \\ &= x + 11x^2 + 11x^3 + x^4 \end{aligned}$$

PROBLEM SET 6.6

6.6.1. **(a)** Find all of the unrestricted partitions of $n = 5$.

(b) How many partitions of $n = 5$ have parts not larger than 3?

(c) How many partitions of $n = 5$ have no more than 3 parts?

Answer

(a) $5, 4+1, 3+2, 3+1+1, 2+2+1, 2+1+1+1, 1+1+1+1+1$, showing that $p(5) = 7$

(b) $3+2, 3+1+1, 2+2+1, 2+1+1+1, 1+1+1+1+1$, showing that $p(5, \leq 3, *) = 5$

(c) $5, 4+1, 3+2, 3+1+1, 2+2+1$, showing that $p(5, *, \leq 3) = 5$.

6.6.3. **(a)** What is the estimate of $p(100) = 190{,}569{,}292$ that is given by formula (6.70)?

(b) The first five terms of the Hardy-Ramanujan series correctly give $p(200) = 3{,}972{,}999{,}029{,}388$. What estimate is given by formula (6.70)?

Answer

(a) 1,999,280,893

(b) 4.10025×10^{12}

6.6.5. Find the generating function for the number of ways to make change for n cents using nickels, dimes, and quarters.

Answer

$$(1 + x^5 + x^{10} + \cdots)(1 + x^{10} + x^{20} + \cdots)(1 + x^{25} + x^{50} + \cdots)$$
$$= \frac{1}{(1 - x^5)(1 - x^{10})(1 - x^{25})}$$

6.6.7. Prove that $p\,(n, \text{odd parts}, =k) = p\,(n - k, \text{ even parts}, \leq k)$.

Answer

Any partition of n that is the sum of k odd integers, $k \leq n$, has the form $a_1, a_2, \ldots, a_j, 1, \ldots, 1$, where $a_j \geq 3$ and $j \leq k$. This corresponds to the partition of $n - k$ of the form $a_1 - 1, a_2 - 1, \ldots, a_j - 1$ that has no more than k even parts. Conversely, a partition $n - k = b_1 + b_2 + \cdots + b_j$ into $j \leq k$ even parts corresponds to a partition $n = (b_1 + 1) + (b_2 + 1) + \cdots + (b_j + 1) + 1 + \cdots + 1$ into exactly k odd parts.

6.6.9. Show that there is always at least one self-conjugate partition of n except when $n = 2$. [*Hint*: See Problem 6.6.8(d).]

Answer

If $n = 2j + 1$, then $2j + 1$ is the partition with one odd part so it corresponds to the self-conjugate partition $(j + 1) + 1 + 1 + \cdots + 1$ with $j + 1$ parts.

If $n = 2j + 2 \geq 4$, then $n = (2j + 1) + 1$ is a partition with two distinct odd parts so it corresponds to the self-conjugate partition $(j + 1) + 2 + 1 + \cdots + 1$ with $j + 1$ parts.

6.6.11. Recall that $p_k(n)$ is the number of partitions of n with exactly k parts, where $p_0(0) = 0$. Prove that

(a) $p_k(n) = p_{k-1}(n-1) + p_k(n-k)$ **(b)** $p_k(n) = \sum_{j=1}^{k} p_j(n-k)$.

Answer

(a) The $p_k(n)$ partitions of n with k parts are of two disjoint types: if the smallest part is $a_k = 1$, subtracting one from this part leaves one of the $p_{k-1}(n-1)$ partitions with $k - 1$ parts; if smallest part is $a_k \geq 2$ subtracting one from each of the k parts leaves one of the $p_k(n-k)$ partitions of $n - k$ still with exactly k parts. The association is reversible, giving the identity.

(b) If $a_1 + a_2 + \cdots + a_j + 1 + \cdots + 1 = n, a_j \geq 2, 1 \leq j \leq k$, is any partition of n with k parts, it corresponds to one of the $p_j(n-k)$ partitions of $n - k$ into j parts, namely $(a_1 - 1) + (a_2 - 1) + \cdots + (a_j - 1) = n - k$. Conversely, if $b_1 + b_2 + \cdots + b_j = n - k, 1 \leq j \leq k$, is one of the $p_j(n-k)$ partitions of $n - k$ with j parts, then $(b_1 + 1) + (b_2 + 1) + \cdots + (b_j + 1) + 1 + \cdots 1 = n$ is a partition of n with k parts.

6.6.13. Prove that the number of unrestricted partitions of n is equal to the number of partitions of $2n$ with exactly n parts; that is, prove that $p(n) = p_n(2n)$.

Answer

If $a_1 + a_2 + \cdots + a_j = n$ is any partition of n, it corresponds to the partition $(a_1 + 1) + (a_2 + 1) + \cdots + (a_j + 1) + 1 + \cdots + 1 = 2n$ with exactly n parts. Conversely, if $b_1 + b_2 + \cdots + b_j + 1 + \cdots + 1 = 2n, b_j \geq 2, 1 \leq j \leq n$, is a partition of $2n$ with n parts, then $(b_1 - 1) + (b_2 - 1) + \cdots + (b_j - 1) = n$ is a partition of n.

6.6.15. Prove that the number of partitions of n into 3 parts is equal to the number of partitions of $2n$ into 3 parts each of size less than n; that is, prove that $p(n, *, = 3) = p(2n, < n, = 3)$.

Answer

Each partition $a_1 + a_2 + a_3 = n$ of n with 3 parts can be uniquely matched to the partition $(n - a_3) + (n - a_2) + (n - a_1) = 2n$ of $2n$ into 3 parts each smaller than n.

6.6.17. **(a)** Find the OGF $g(x)$ of the sequence $p(n,$ distinct powers of 2, $*)$.
(b) What does part (a) say about the base two representation of an integer?

Answer

(a) The OGF is

$$g(x) = \prod_{j\geq 0}(1 + x^{2^j}) = (1+x)(1+x^2)(1+x^4)(1+x^8)\cdots$$
$$= \frac{(1-x^2)(1-x^4)(1-x^8)\cdots}{(1-x)(1-x^2)(1-x^4)(1-x^8)\cdots} = \frac{1}{1-x} = 1 + x + x^2 + \cdots$$

(b) There is just one way to represent n as a sum of distinct powers of 2; that is, the binary representation of any integer is unique.

6.6.19. Use a computer algebra system (CAS) to verify that

$$\sum_{n=0}^{\infty} p(n, *, \leq 3)x^n = \frac{1}{(1-x)}\frac{1}{(1-x^2)}\frac{1}{(1-x^3)} = \frac{1}{6}\frac{1}{(1-x)^3}$$
$$+ \frac{1}{4}\frac{1}{(1-x)^2} + \frac{1}{4}\frac{1}{(1-x^2)} + \frac{1}{3}\frac{1}{(1-x^3)}$$

Answer

For example, in *Maple*, the commands

```
f:=(1-x)*(1-x^2)*(1-x^3);
g:=1/(1-x)^3/6+1/(1-x)^2/4+1/(1-x^2)/4+1/(1-x^3)/3;
simplify (f(x)*g(x));
```

return a 1.

6.6.21. Consider the partitions of n into distinct parts. If n is not a generalized pentagonal number, then the proof of Theorem 6.52 gave a pairing of partitions with evenly many parts to partitions with an odd number of parts. However, if n is a pentagonal number, all but one exceptional partition can be paired. Using Ferrers diagrams, show

(a) the pairings for $n = 8$

(b) the pairings and the exceptional partition for $n = 5$

(c) the pairings and the exceptional partition for $n = 7$

Answer

(a) There are 6 partitions, each paired.

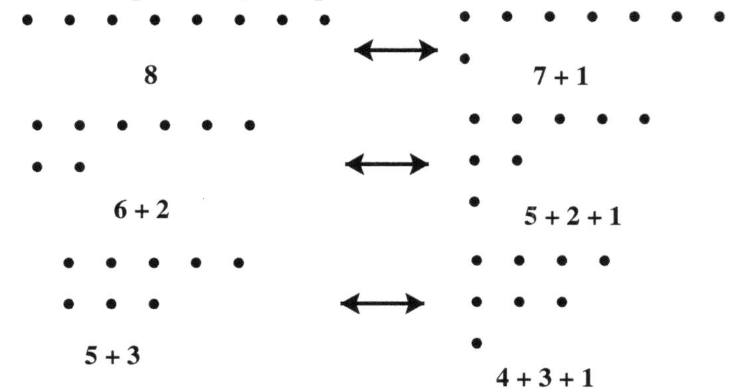

(b) There are 3 partitions, with two paired and one exception.

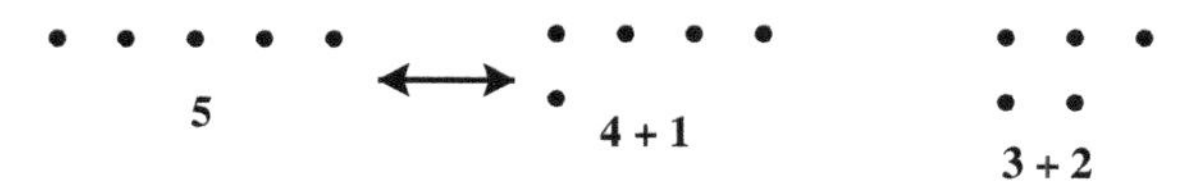

(c) There are 5 partitions, with 4 paired and one exception.

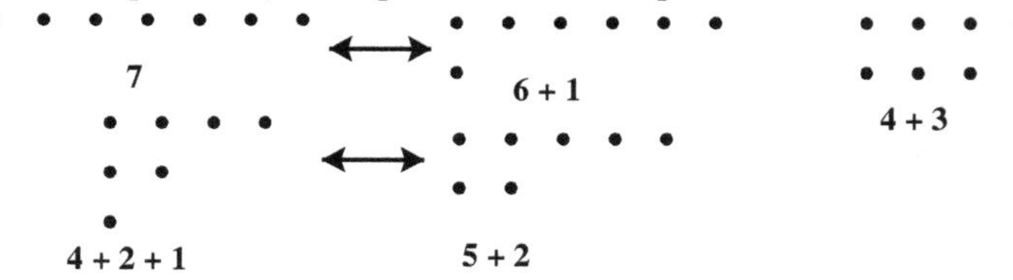

6.6.23. According to Theorem 6.51, the generating function for the number of partitions of the form $a_1 + a_2 + a_3 = n$ with $a_1 \geq a_2 \geq a_3 \geq 1$ is $\dfrac{x^3}{(1-x)(1-x^2)(1-x^3)}$.

The parts a_1, a_2, a_3 will be the sides of an integer sided triangle of perimeter n unless $a_1 \geq a_2 + a_3$, since this would violate the triangle inequality.

(a) Show that the generating function for the number of partitions of n with three parts and for which $a_1 \geq a_2 + a_3$ is $\dfrac{x^4}{(1-x)(1-x^2)(1-x^4)}$.

(b) Show that the generating function for T_n, the number of noncongruent triangles of perimeter n with integer sides, is

$$\frac{x^3}{(1-x)(1-x^2)(1-x^3)} - \frac{x^4}{(1-x)(1-x^2)(1-x^4)}$$

$$= \frac{x^3}{(1-x^2)(1-x^3)(1-x^4)}$$

Answer

(a) Let the integer variables x and y be defined by $a_1 = a_2 + x$ and $a_2 = a_3 + y$, so $x \geq 0$, $y \geq 0$ and $n = a_1 + a_2 + a_3 = 2a_2 + x + a_3 = 2(a_3 + y) + x + a_3 = x + 2y + 3a_3$. Note that the OGF for the number of solutions of $x + 2y + 3a_3 = n$ with $x \geq 0, y \geq 0, a_3 \geq 1$ is $\frac{x^3}{(1-x)(1-x^2)(1-x^3)}$. If $a_1 \geq a_2 + a_3$, or $a_2 + x \geq a_2 + a_3$, we see that $x = a_3 + z$ for some $z \geq 0$. This gives the equation $n = x + 2y + 3a_3 = z + 2y + 4a_3$, where $z \geq 0$, $y \geq 0$, and $a_3 \geq 1$. The OGF for the number of solutions is then $\frac{x^4}{(1-x)(1-x^2)(1-x^4)}$.

(b) Subtraction of the OGFs gives

$$\frac{x^3}{(1-x)(1-x^2)(1-x^3)} - \frac{x^4}{(1-x)(1-x^2)(1-x^4)}$$

$$= \frac{x^3}{(1-x)(1-x^2)}\left(\frac{1}{1-x^3} - \frac{x}{1-x^4}\right)$$

$$= \frac{x^3}{(1-x)(1-x^2)(1-x^3)(1-x^4)}(1 - x^4 - x + x^4)$$

$$= \frac{x^3}{(1-x^2)(1-x^3)(1-x^4)}$$

PROBLEM SET 6.7

6.7.1. What is the number of mountain ranges across the Isthmus of Panama (sea level to sea level) that can be drawn with n upstrokes and n downstrokes? The elevation gain or loss of each stroke is the same, and no pass between adjacent mountains is below sea level.

Answer

A mountain range corresponds to a sequence of n positive ones(upstrokes) and n negative ones (downstrokes) with positive partial sums, so there are C_n mountain ranges.

6.7.3. Raffle tickets are \$5 each. What is the probability that 12 people can buy a ticket, where 6 people have a \$5 bill and the other 6 people have a \$10 bill and enough change is on hand for each successive transaction? At the beginning of the ticket sales, there is no money in the collection box.

Answer

Change will always be available if at least as many people with \$5 bills are in front of those needing change. There are 12! ways for the people to line up, of which $(6!)^2\, C_6$ ways have no problem making change. This gives the probability $\frac{6!^2 C_6}{12!} = \frac{1}{6+1} = \frac{1}{7}$.

6.7.5. The sum of the terms of the sequence $(-1, -1, 1, 1, -1, -1, 1, 1, 1)$ is 1. Graph the path generated by the sequence, and use it to identify the unique cyclic shift of the sequence with all positive partial sums.

Answer
Since -1 corresponds to a move north, and $+1$ to a block east, the sequence generates the path from A to A' shown below. The point B is the uppermost point of the path farthest from the 45° line through A. It is the beginning point of the cyclic shift $(1, 1, 1 -1, - 1, 1, 1, -1, -1)$ with all positive partial sums.

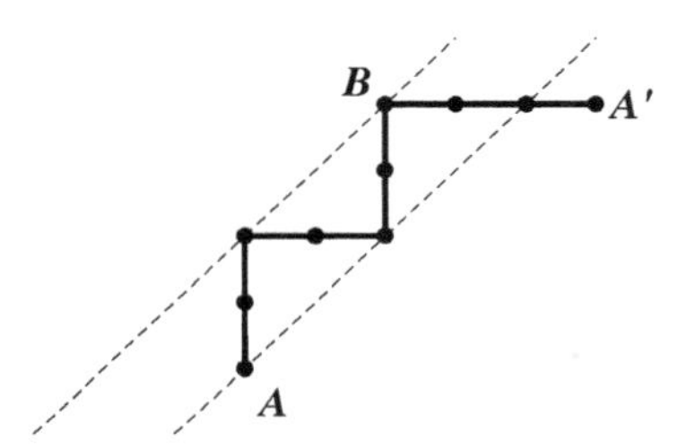

6.7.7. Given any sequence of integers $(a_1, a_2, \dots, a_n)$ for which $a_1 + a_2 + \cdots + a_n = 1$, prove there is a unique cyclic shift with all positive partial sums.

Answer
The given sequence corresponds to a path from A to A' with point A' one unit to the right of the 45° line through A. Any cyclic shift is a path of n steps along the extended path created by translates of the starting path by the vector A to A'. Only the cyclic shift beginning at the uppermost point B of the 45° line for which all further points lie below the 45° line gives a path with all positive partial sums.

6.7.9. Draw all of the triangulations determined by the diagonals of a convex hexagonal region that do not intersect at an interior point of the region.

Answer
The $C_4 = 14$ triangulations are

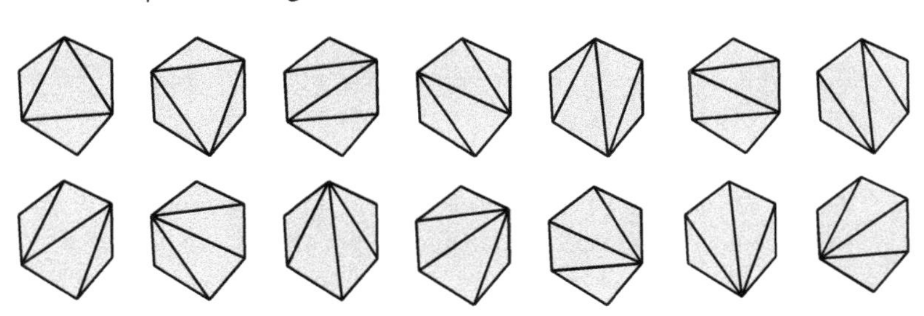

6.7.11. Prove the statement of Catalan 2; that is, prove that C_n is the number of ways of connecting n points on the x-axis by noncrossing arcs in the upper half plane, such that if two arcs share an endpoint p, then p is a left endpoint of both arcs.

Answer

Use strong mathematical induction, assuming there are C_k arrangements of arcs on any row of k dots for which $1 \leq k \leq n$. Now consider arrangements on a row of dots numbered 1 through $n + 1$. There are C_n arrangements with no arc originating on dot 1, since these are just the arrangements on dots 2 through $n + 1$. Next suppose there is at least one arc originating at dot 1, with the longest arc ending at dot $k + 1$. There are C_k such arrangements, because these are just the arrangements on dots 1 through k together with the arc from 1 to $k + 1$. There are also C_{n-k} arrangements on the $n - k$ dots numbered $k + 2$ through $n + 1$. This means there are $C_k\, C_{n-k}$ arrangements with the longest arc from dot 1 to dot $k + 1$. Together with the C_n arrangements with no arc at dot 1, there are $\sum_{k=0}^{n} C_k C_{n-k}$ arrangements on $n + 1$ dots. But this sum is C_{n+1} by formula (6.77), which completes the induction step.

6.7.13. Using the labeling algorithm described in Example 6.57, draw the triangulation of the polygonal region that corresponds to the parenthesis placement $(((v_0v_1)v_2)(((v_3v_4)v_5)v_6))$.

Answer

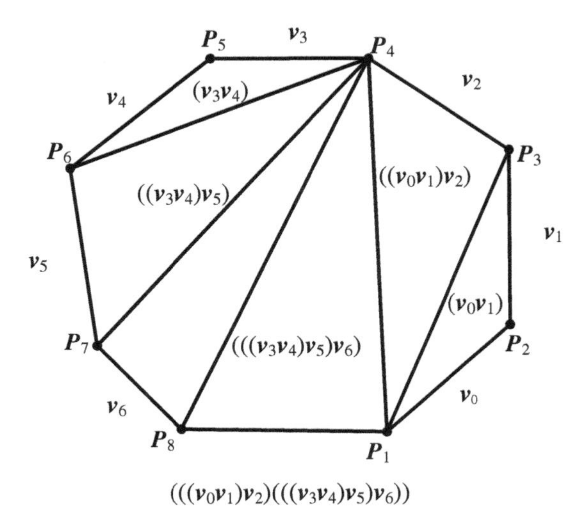

6.7.15. Example 6.57 can be modified to show that the super Catalan numbers of Problem 6.7.14 give the number of arbitrary bracketings of a list of symbols. For example, the bracketed list $(((v_0v_1v_2)(v_3v_4))v_5v_6)$ corresponds to the edge and diagonal labeling of the dissection shown below.

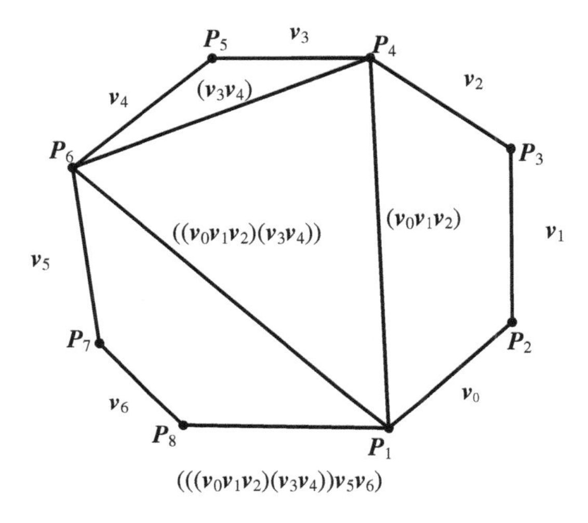

(a) Give the bracketed list of symbols that corresponds to this dissection of a polygonal region.

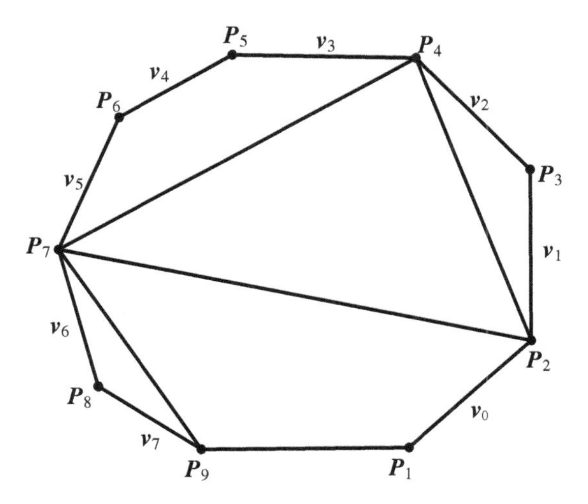

(b) Draw and label the dissected polygonal region that corresponds to this bracketed list of symbols $(v_0(v_1v_2)((v_3v_4v_5)v_6)v_7v_8)$.

Answer

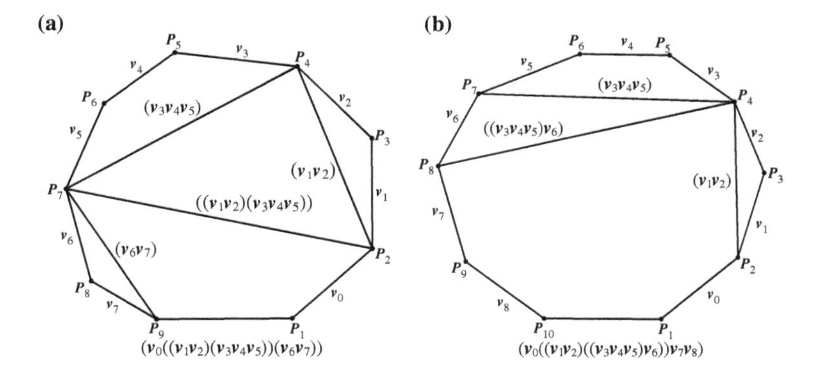

PROBLEM SET 6.8

6.8.1. Find the quartic polynomial $p(n)$ that gives the maximum number of interior regions within a disk formed by all of the chords drawn between pairs of the n points on the circular boundary of the disk. For example, $p(5) = 16$, as shown in this diagram. Notice that $p(0) = 1$, $p(1) = 1$, and $p(2) = 2$.

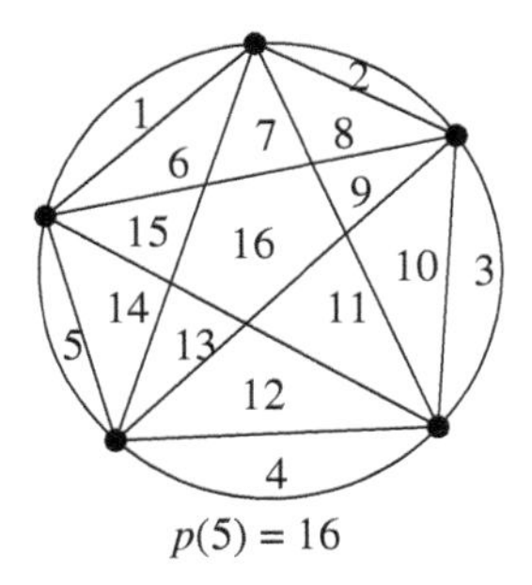

$p(5) = 16$

Answer

Diagrams show that the number of regions is the sequence 1, 1, 2, 4, 8, 16, 31. … The difference table is therefore

1		1		2		4		8		16		31
	0		1		2		4		8		15	
		1		1		2		4		7		
			0		1		2		3			
				1		1		1				
					0		0					

By Theorem 6.3, $p(n) = \binom{n}{0} + \binom{n}{2} + \binom{n}{4}$. Then using the Pascal's identity, $p(n) = \binom{n-1}{0} + \binom{n-1}{1} + \binom{n-1}{2} + \binom{n-1}{3} + \binom{n-1}{4}$ for $n \geq 1$, which is the sum of the first five entries in row $n-1$ of Pascal's triangle. (Compare to Problem 7.3.3, which shows the assumption that the answer is a quartic polynomial is unnecessary.)

6.8.3. Let $q(0) = 0$ and $q(n) = q(n-1) + s_n$, $n \geq 1$, where s_n is the sum of the base ten digits of n. For example, $q(1) = q(0) + 1 = 1$, $q(2) = q(1) + 2 = 3$, and $q(3) = q(2) + 3 = 6$.

(a) Show that $q(n) = \binom{n+1}{2}$ for $0 \leq n \leq 8$.

(b) Is $q(n) = \binom{n+1}{2}$ for all $n \geq 0$?

Answer

(a)

$$q(4) = 6 + 4 = 10 = \binom{4+1}{2}, \; q(5) = 10 + 5 = 15 = \binom{5+1}{2},$$

$$q(6) = 15 + 6 = 21 = \binom{6+1}{2}, \; q(7) = 21 + 7 = 28 = \binom{7+1}{2},$$

$$\text{and } q(8) = 28 + 8 = 36 = \binom{8+1}{2}.$$

(b) No. For example,

$$q(9) = 36 + 9 = 45 = \binom{10}{2} \text{ but } q(10) = 45 + 1 = 46 \neq 55 = \binom{11}{2}.$$

6.8.5. The *associated Stirling numbers of the second kind* $\left\{\!\!\left\{ {k \atop j} \right\}\!\!\right\}$ give the number of partitions of $[k]$ into j subsets that each contain two or more members.

(a) Show that

$$\left\{\!\!\left\{ {k+1 \atop j} \right\}\!\!\right\} = j \left\{\!\!\left\{ {k \atop j} \right\}\!\!\right\} + k \left\{\!\!\left\{ {k-1 \atop j-1} \right\}\!\!\right\}, \; k \geq 3.$$

(b) Extend the triangular table of the associated Stirling numbers of the second kind shown below to include the entries for k and j with $1 \le 2j \le k \le 9$.

	$j = 1$	2	3
$k = 1$	0	0	
2	1	0	
3	1	0	
4	1	3	0
5	1	10	0

(A spreadsheet is helpful.)

Answer

(a) There are $j \left\{\!\!\left\{ {k \atop j} \right\}\!\!\right\}$ partitions for which element $k + 1$ is in a subset with 3 or more members: first, in $\left\{\!\!\left\{ {k \atop j} \right\}\!\!\right\}$ ways, partition $[k]$ into j subsets each with 2 or more members, and then add element $k + 1$ to one of these subsets in j ways. There are also $k \left\{\!\!\left\{ {k-1 \atop j-1} \right\}\!\!\right\}$ partitions for which element $k + 1$ is in a doubleton subset: choose the second member of the doubleton in k ways and then partition the remaining $k - 1$ elements into subsets with 2 or more members in $\left\{\!\!\left\{ {k-1 \atop j-1} \right\}\!\!\right\}$ ways.

(b)

	$j = 1$	2	3	4
$k = 1$	0	0		
2	1	0		
3	1	0		
4	1	3	0	
5	1	10	0	
6	1	25	15	0
7	1	56	105	0
8	1	119	490	105
9	1	246	1918	1260

6.8.7. Use the result of Problem 6.8.6(b) and Theorem 6.24(c) to prove that

$$\sum_{k=1}^{m} (-1)^{m+k} \left\{ {m \atop k} \right\} \left[{k+1 \atop 2} \right] = m.$$

Answer

Let $g_n = (-1)^n n$. Therefore,

$$f_n = \sum_{j=0}^{n} (-1)^j \begin{bmatrix} n \\ j \end{bmatrix} g_j = \sum_{j=0}^{n} (-1)^j \begin{bmatrix} n \\ j \end{bmatrix} (-1)^j j = \sum_{j=0}^{n} j \begin{bmatrix} n \\ j \end{bmatrix} = \begin{bmatrix} n+1 \\ 2 \end{bmatrix}$$

and $(-1)^m m = g_m = \sum_{k=1}^{m} (-1)^k \left\{ \begin{matrix} m \\ k \end{matrix} \right\} f_k = \sum_{k=1}^{m} (-1)^k \left\{ \begin{matrix} m \\ k \end{matrix} \right\} \begin{bmatrix} k+1 \\ 2 \end{bmatrix}$.

6.8.9. Let $S_k = \sum_{n=0}^{k} a_n b_n$ and $B_k = \sum_{n=0}^{k} b_n$. Prove that

$S_k = a_k B_k - \sum_{n=0}^{k-1} B_n \left(a_{n+1} - a_n\right)$, called the *Abel transformation.*

Answer

Using partial summation, where $b_n = \Delta B_{n-1}$ for all $n > 1$,

$$S_k = a_0 b_0 + a_k b_k + \sum_{n=1}^{k-1} a_n \Delta B_{n-1} = a_0 b_0 + a_k b_k + a_n B_{n-1} \Big|_{n=1}^{n=k} - \sum_{n=1}^{k-1} E B_{n-1} \Delta a_n$$

$$= a_0 B_0 + a_k \left(B_k - B_{k-1}\right) + a_k B_{k-1} - a_1 B_0 - \sum_{n=1}^{k-1} B_n \left(a_{n+1} - a_n\right)$$

$$= a_k B_k - B_0 \left(a_1 - a_0\right) - \sum_{n=1}^{k-1} B_n \left(a_{n+1} - a_n\right)$$

$$= a_k B_k - \sum_{n=0}^{k-1} B_n \left(a_{n+1} - a_n\right)$$

6.8.11. **(a)** Give the 10 partitions with no more than two parts and of size at most three, including the null partition 0.

(b) Let j and k be positive integers. Prove there are $\binom{j+k}{k}$ partitions with no more than k parts of size at most j.

Answer

(a) These are the partitions whose Ferrers diagrams are a sublattice of the 2×3 lattice. The partitions, shown by the black dots, are:

○ ○ ○   ● ○ ○   ● ○ ○   ● ● ○   ● ● ○
○ ○ ○   ○ ○ ○   ● ○ ○   ○ ○ ○   ● ○ ○

● ● ○   ● ● ●   ● ● ●   ● ● ●   ● ● ●
● ● ○   ○ ○ ○   ● ○ ○   ● ● ○   ● ● ●

Thus the ten partitions are 0, 1, 1 + 1, 2, 2 + 1, 2 + 2, 3, 3 + 1, 3 + 2, and 3 + 3.

(b) Any Ferrers diagram of the type shown in part (a), when reflected over the 45° upward diagonal, can be viewed as the graph of a nondecreasing function from $[k]$ into $\{0, 1, 2, \ldots, j\}$ as shown here:

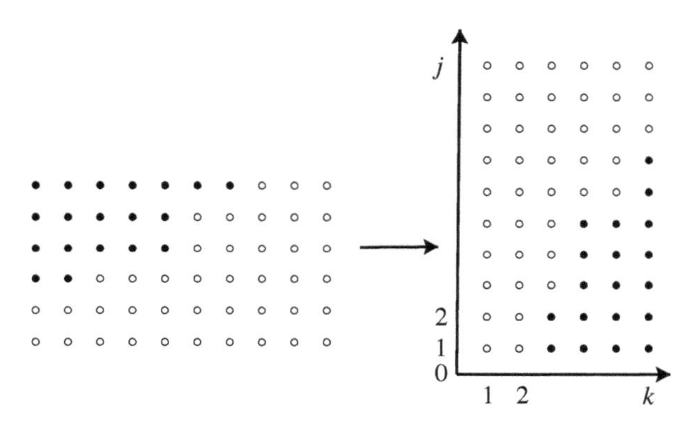

We have already seen (Theorem 2.48) that the number of such functions, and therefore the number of partitions, is given by $\left(\!\binom{j+1}{k}\!\right) = \binom{j+k}{k}$.

PART II

TWO ADDITIONAL TOPICS IN ENUMERATION

7

LINEAR SPACES AND RECURRENCE SEQUENCES

PROBLEM SET 7.2

In Problems 7.2.1 through 7.2.6, let

$$t_{n+2} = 5t_{n+1} - 6t_n; t_0 = 0, t_1 = 1$$
$$u_{n+2} = 3u_{n+1} - 2u_n; u_0 = 0, u_1 = 1$$
$$v_{n+3} = 3v_{n+2} + 4v_{n+1} - 12v_n; v_0 = 1, v_1 = -1, v_2 = 9$$

7.2.1. Find the spectrum (multiset of eigenvalues) of each of these sequences.
(a) t_n **(b)** u_n **(c)** v_n

Answer
(a) $\{2, 3\}$ **(b)** $\{1, 2\}$ **(c)** $\{2, -2, 3\}$

7.2.3. Find the spectrum, GPS, and order of these sequences.
(a) t_{2n} **(b)** u_{3n} **(c)** t_n^2 **(d)** u_{2n}^2

Answer
(a) $\{4, 9\}$, $t_{2n} = 9^n - 4^n$, order 2
(b) $\{1, 8\}$, $u_{3n} = 8^n - 1$, order 2

Solutions Manual to Accompany Combinatorial Reasoning: An Introduction to the Art of Counting, First Edition. Duane DeTemple and William Webb.

(c) $\{4, 6, 9\}, t_n^2 = 9^n - 2 \cdot 6^n + 4^n$, order 3

(d) $\{1, 4, 16\}, u_{2n}^2 = 16^n - 2 \cdot 4^n + 1$, order 3

7.2.5. **(a)** What upper bounds for the orders of v_n^2 and $t_{2n} + v_n^2$ are given by Theorems 7.10, 7.11 and 7.13?

(b) What are the spectra of v_n^2 and $t_{2n} + v_n^2$? What bounds do these multisets give for the order of the sequences?

Answer

(a) $\mathcal{O}\left(v_n^2\right) \leq \binom{3+2-1}{2} = 6$ and $\mathcal{O}\left(t_{2n} + v_n^2\right) \leq 2 + 6 = 8$

(b) $\mathcal{E}\left(v_n^2\right) = \{4, 9, -4, 6, -6\}$, so that $\mathcal{O}\left(v_n^2\right) \leq 5$
$\mathcal{E}\left(t_{2n} + v_n^2\right) = \{4, 9, -4, 6, -6\}$, so that $\mathcal{O}\left(t_{2n} + v_n^2\right) \leq 5$

7.2.7. Find the spectrum and order of the sequence $h_n = 2^n + n^2$.

Answer

$\mathcal{E}\left(h_n\right) = \{2, 1, 1, 1\}$, so that $\mathcal{O}\left(h_n\right) = 4$

7.2.9. Let $C(E) = \sum_{j=0}^{k} a_j E^{k-j}$ annihilate the sequence h_n and have the spectrum $\mathcal{E}_C = \left\{r_1 \cdot \alpha_1, r_2 \cdot \alpha_2, \ldots, r_m \cdot \alpha_m\right\}$.

(a) For any $\beta \neq 0$, show that the operator $C_\beta(E) = \beta^k \sum_{j=0}^{k} a_j \left(\frac{E}{\beta}\right)^{k-j}$ annihilates the sequence $\beta^n h_n$.

(b) What is the spectrum of the sequence $\beta^n h_n$?

Answer

(a) $C_\beta(E)\beta^n h_n = \beta^k \left(\sum_{j=0}^{k} a_j \left(\frac{E}{\beta}\right)^{k-j}\right)\beta^n h_n = \sum_{j=0}^{k}\left(a_j \beta^j E^{k-j}\right)\beta^n h_n =$
$\sum_{j=0}^{k} a_j \beta^j \beta^{n+k-j} h_{n+k-j} = \beta^{n+k} \sum_{j=0}^{k} a_j h_{n+k-j} = \beta^{n+k} C(E)\, h_n = 0$

(b) Since $C_\beta(x) = \beta^k \sum_{j=0}^{k} a_j \left(\frac{x}{\beta}\right)^{k-j} = \beta^k C\left(\frac{x}{\beta}\right)$, any eigenvalue α_i of $C(E)$ corresponds to the eigenvalue $\beta\alpha_i$ of $C_\beta(E)$ with the same multiplicity r_i. That is, $\mathcal{E}_{C_\beta} = \left\{r_1 \cdot (\beta\alpha_1), r_2 \cdot (\beta\alpha_2), \ldots, r_m \cdot (\beta\alpha_m)\right\}$. This also follows from $\mathcal{E}_{CD} = \mathcal{E}_C \otimes \mathcal{E}_D = \mathcal{E}_C \otimes \{\beta\}$, where $D(E) = E - \beta$ and $\otimes$ denotes the set of products.

7.2.11. Show that the generating function of the sequence P_nF_n discussed in Example 7.6 is

$$f_{PF}(x) = \frac{x\left(1-x^2\right)}{1-2x-7x^2-2x^3+x^4}$$

Answer

The form of the OGF is

$$\sum_{n\geq 0} P_nF_nx^n = 0 + x + 2x^2 + 10x^3 + 36x^4 + \cdots = \frac{b_0 + b_1x + b_2x^2 + b_3x^3}{1-2x-7x^2-2x^3+x^4}$$

Therefore

$$b_0 + b_1x + b_2x^2 + b_3x^3 = \left(x + 2x^2 + 10x^3 + \cdots\right) \times \left(1-2x-7x^2-2x^3+x^4\right) = x - x^3$$

7.2.13. If $u_n = \left(c_0 + c_1 n\right)\alpha^n,\ c_1 \neq 0$, determine a homogeneous recurrence relation of order $t+1$ that is satisfied by $u_n^t = \left(c_0 + c_1 n\right)^t \alpha^{nt}$.

Answer

Since $u_n^t = \left(c_0 + c_1 n\right)^t \alpha^{nt} = p(n)\left(\alpha^t\right)^n$ where $p(n)$ is a polynomial in n of degree t, we see that u_n^t is annihilated by the operator $\left(E - \alpha^t\right)^{t+1}$.

PROBLEM SET 7.3

7.3.1. Find a polynomial $f(E)$ that annihilates h_n given

(a) $h_{n+2} - 3h_{n+1} + 7h_n = (-5)^n$

(b) $h_{n+2} - 3h_n = n^2$

(c) $h_{n+3} - h_{n+2} + 2h_n = 3 \cdot 2^n + 7$

(d) $h_{n+2} + h_{n+1} - 4h_n = 5n + 2\left(-1^n\right) + 3^n$

(e) $h_{n+2} - h_{n+1} - h_n = 3F_n$

Answer

(a) $(E+5)\left(E^2 - 3E + 7\right)$

(b) $(E-1)^3\left(E^2 - 3\right)$

(c) $(E-1)(E-2)\left(E^3 - E^2 + 2\right)$

(d) $(E-1)^2(E+1)(E-3)\left(E^2 + E - 4\right)$

(e) $\left(E^2 - E - 1\right)^2$

7.3.3. Let g_n denote the number of regions formed inside a circle when n points are placed on the circumference and all chords are drawn, where it is assumed that no three chords intersect at a common point interior to the circle. Thus, $g_0 = 1, g_1 = 1$, and $g_2 = 2$.

(a) Obtain a nonhomogeneous recursion relation for g_n.

(b) Use the recurrence to compute g_n for $n = 2, 3, 4, 5, 6$.

(c) Determine the formula for g_n using Theorem 7.17.

(d) Determine the formula for g_n by summing the telescoping sum $\sum_{k=0}^{n-1} (g_{k+1} - g_k)$.

Answer

(a) Suppose there are n points on a circle, with g_n regions formed by the chords. An additional point forms a new region for each of the $\binom{n}{3}$ triples of the original n points, plus n additional regions which have a vertex at the new point. Thus, $g_{n+1} = g_n + \binom{n}{3} + n$.

(b) $g_3 = 2 + 2 + 0 = 4, g_4 = 4 + 3 + \binom{3}{3} = 8, g_5 = 8 + 4 + \binom{4}{3} = 16, g_6 = 16 + 5 + \binom{5}{3} = 31, g_7 = 31 + 6 + \binom{6}{3} = 57$

(c) Since the nonhomogeneous term of the recurrence is a cubic polynomial, an annihilating polynomial for g_n is $(E - 1)^5$, so g_{n+1} is a polynomial of degree at most 4, say $g_{n+1} = c_0\binom{n}{0} + c_1\binom{n}{1} + c_2\binom{n}{2} + c_3\binom{n}{3} + c_4\binom{n}{4}$. Then $1 = g_1 = c_0, 2 = g_2 = 1 + c_1$ and so on, showing that

$$g_{n+1} = \binom{n}{0} + \binom{n}{1} + \binom{n}{2} + \binom{n}{3} + \binom{n}{4}$$

(d) By the telescopic property and the hockey stick identities,

$$g_{n+1} = 1 + \sum_{k=0}^{n} (g_{k+1} - g_k) = 1 + \sum_{k=0}^{n} \binom{k}{1} + \sum_{k=0}^{n} \binom{k}{3}$$

$$= 1 + \binom{n+1}{2} + \binom{n+1}{4} = \binom{n}{0} + \binom{n}{1} + \binom{n}{2} + \binom{n}{3} + \binom{n}{4}.$$

(Compare with Problem 6.8.1.)

7.3.5. Suppose that $C(E)\,u_n = 0$ and $D(E)\,v_n = 0$. Find an annihilating polynomial for h_n when

(a) $C(E)\,h_n = u_n$

(b) $C(E)\,h_n = u_n + nv_n$

Answer

(a) $C(E)^2$

(b) $C(E)^2\,D(E)^2$

7.3.7. Let h_n count the number strings of the digits 0, 1, 2, … ,9 of length n, where a digit following a prime digit must also be a prime digit, a 4, 6, or 8 can only be followed by a 0, and a 9 can only be followed by a 9. For example, 11409 and 60172 are permissible strings of length 5, but 01754 and 80599 are not allowed.

(a) Obtain a recurrence for h_n.

(b) Solve the recurrence for h_n.

Answer

(a) There are 4^n strings with all primes from the set of prime digits $\{2, 3, 5, 7\}$, and there is one string of all 9s. There are also $2h_{n-1}$ strings that begin with 0 or 1, and another $3h_{n-2}$ strings that begin with 40, 60, or 80, where $h_0 = 1$. This gives the recurrence relation $h_n = 2h_{n-1} + 3h_{n-2} + 4^n + 1$.

(b) $(E-4)(E-1)\left(E^2-2E-3\right) = (E-4)(E-1)(E+1)(E-3)$ is an annihilating operator, so $h_n = c_1 + c_2 4^n + c_3(-1)^n + c_4 3^n$. The initial conditions are $h_0 = 1, h_1 = 10, h_2 = 40, h_3 = 175$. Solving the system

$$\begin{aligned} c_1 + c_2 + c_3 + c_4 &= 1 \\ c_1 + 4c_2 - c_3 + 3c_4 &= 10 \\ c_1 + 16c_2 + c_3 + 9c_4 &= 40 \\ c_1 + 64c_2 - c_3 + 27c_4 &= 175 \end{aligned}$$

gives us $h_n = -\dfrac{1}{4} + \dfrac{16}{5}\cdot 4^n - \dfrac{33}{40}(-1)^n - \dfrac{9}{8}\cdot 3^n$.

7.3.9. What recurrence is satisfied by both of the sequences x_n and y_n if $x_n = x_{n-1} + x_{n-2} - y_{n-1}, y_n = 2x_{n-2} + 3y_{n-2}$?

Answer

First rewrite the system in the standard form

$$\begin{aligned} x_{n+2} - x_{n+1} - x_n + y_{n+1} &= 0 \\ -2x_n + y_{n+2} - 3y_n &= 0 \end{aligned}$$

Therefore $\det\begin{bmatrix} E^2-E-1 & E \\ -2 & E^2-3 \end{bmatrix} = E^4 - E^3 - 4E^2 + 5E + 3$ annihilates both x_n and y_n. That is, both sequences satisfy $h_{n+4} - h_{n+3} - 4h_{n+2} + 5h_{n+1} + 3h_n = 0$.

7.3.11. How many block-walking paths of length n move along blocks to the east, west, or north, never returning to the same intersection a second time? For example, NEENWW and WNNEEN are permissible walks of length 6, but NWEEWN is not allowed since the WE pair returns the path to a previously occupied position, as does the EW pair. [*Hint*: Let u_n denote the number of paths of length n that begin with a block to the north and let v_n denote the number of paths of length n that begin with a block to the east or, equivalently, with a block to the west.]

Answer
If the first block of a path of length $n+1$ is to the north, then next block can be north, east, or west, so $u_{n+1} = u_n + 2v_n$. If the first block is to the east, the next block must be north or east, so $v_{n+1} = u_n + v_n$. Thus the polynomial operator $\det\begin{bmatrix} E-1 & -2 \\ -1 & E-1 \end{bmatrix} = (E-1)^2 - 2 = E^2 - 2E - 1$ annihilates both sequences. Since this is the Pell operator, we have the form $u_n = c_1 P_{n-1} + c_2 P_n$, where P_n denotes the Pell sequence 0, 1, 2, 5, 12, 29, ... Since $u_1 = 1$and $v_1 = 1$ we also have $u_2 = u_1 + 2v_1 = 3$and $v_2 = u_1 + v_1 = 2$. These conditions show that $u_n = P_{n-1} + P_n$ and $v_n = P_n$. The total number of paths of length n is $u_n + 2v_n = u_{n+1} = P_n + P_{n+1}$.

PROBLEM SET 7.4

Unless stated otherwise, use the operator method to prove identities and express sums in closed form. Recall that $\sum_j$ and $\sum$ both denote a summation over all integers j.

7.4.1. Prove that

(a) $F_n^2 + F_{n+1}^2 = F_{2n+1}$

(b) $F_{n+1}^2 - F_{n-1}^2 = F_{2n}$

(c) $F_{n+1}F_{n-1} - F_n^2 = (-1)^n$ (Simson's identity)

(d) $F_n F_{n+1} = \frac{1}{5}\left(2F_{2n} + F_{2n+1} - (-1)^n\right)$

Answer
The spectra of all of the terms such as $F_n^2, F_nF_{n+1}, F_{2n}$, and $(-1)^n$ that appear in the formulas are included in the spectrum $\{\varphi^2, \hat{\varphi}^2, -1\}$,

so all of the terms are annihilated by $(E+1)\left(E-\varphi^2\right)\left(E-\hat{\varphi}^2\right) = (E+1)\left(E^2-3E+1\right)$. Since each equation (a), (b), (c) and (d) holds for $n = 0$, 1, and 2, each identity is proved.

7.4.3. Prove that $F_{m+n} = \frac{1}{2}\left(F_m L_n + F_n L_m\right)$.

Answer

View m as a fixed parameter. Then since the Fibonacci operator $f(E) = E^2 - E - 1$ annihilates F_{m+n} for each m, there are constants a_m and b_m, dependent on m, for which $F_{m+n} = a_m L_n + b_m F_n$. For $n = 0$, we get $F_m = a_m L_0 + 0 = 2a_m$ so $a_m = \frac{F_m}{2}$ and $F_{m+n} = \frac{F_m L_n}{2} + b_m F_n$. Switching m with n gives $F_{m+n} = F_{n+m} = \frac{F_n L_m}{2} + b_n F_m$, which shows that the coefficients of F_n is $b_m = \frac{L_m}{2}$.

Alternate answer

Let E_n denote the successor operator on the free variable n. Then $f\left(E_n\right) = E_n^2 - E_n - 1$ annihilates both of the sequences

$$s_{n,0} = F_n - \frac{1}{2}\left(F_0 L_n + F_n L_0\right) \text{ and } s_{n,1} = F_n - \frac{1}{2}\left(F_1 L_n + F_n L_1\right)$$

and both sequences are 0 for $n = 0,1$. Therefore, both are the 0 sequences. Now let E_m denote the successor operator on the free variable m. Then, for any n, the operator $f\left(E_m\right) = E_m^2 - E_m - 1$ annihilates the sequence

$$s_{n,m} = F_{n+m} - \frac{1}{2}\left(F_m L_n + F_n L_m\right)$$

But $s_{n,0} = s_{n,1} = 0$, so $s_{n,m} = 0$ for all m and n.

7.4.5. Let $t_{n+2} = 5t_{n+1} - 6t_n, t_0 = 0, t_1 = 1$.

(a) What is the spectrum associated with the sum $s_n = \sum_{i=0}^{n} t_i$?

(b) Express s_n as a GPS.

(c) Give another form of s_n as an expression involving the terms of the sequence t_n.

Answer

(a) $\{1, 2, 3\}$

(b) $s_n = c_0 + c_1 2^n + c_2 3^n$, so using the initial values $s_0 = 0, s_1 = 1, s_2 = 6$ gives $s_n = \frac{1}{2} - 2 \cdot 2^n + \frac{3}{2} \cdot 3^n = \frac{1}{2}\left(1 - 2^{n+2} + 3^{n+1}\right)$

(c) $s_n = d_0 + d_1 t_n + d_2 t_{n+1}$, so the initial values $s_0 = 0, s_1 = 1, s_2 = 6$ gives $s_n = \frac{1}{2} + 3t_n - \frac{1}{2}t_{n+1}$

7.4.7. Let $h_{n+2} - ah_{n+1} - bh_n = 0$ where $h_0 = 0, h_1 = 1$ and $a + b \neq 1$.

(a) Show that 1 is not an eigenvalue of h_n.

(b) Find a closed form of the sum $s_n = \sum_{i=0}^{n} h_i$.

Answer

(a) If 1 and α were the eigenvalues for the operator $C(E) = E^2 - aE - b$, this would mean $C(x) = x^2 - ax - b = (x-1)(x-\alpha) = x^2 - (1+\alpha)x + \alpha$, so $\alpha = -b$ and $1 + \alpha = a$. This shows that $a + b = 1$, a contradiction.

(b) Since 1 is not an eigenvalue of the sequence h_n, the sum s_n has the form $s_n = c_0 + c_1 h_n + c_2 h_{n+1}$. Using the initial values

$$h_0 = 0, h_1 = 1, h_2 = a, h_3 = a^2+b, h_4 = a^3 + 2ab$$

$$s_0 = 0, s_1 = 1, s_2 = a + 1, s_3 = a^2 + a + b + 1$$

to evaluate the constants, we get the closed form $s_n = \dfrac{bh_n + h_{n+1} - 1}{a + b - 1}$.

7.4.9. Prove that

(a) $\sum_j \binom{n}{j} F_{1+j} = F_{2n+1}, n \geq 0$

(b) $\sum_j \binom{n}{j} F_{2+j} = F_{2n+2}, n \geq 0$

(c) $\sum_j \binom{n}{j} F_{m+j} = F_{2n+m}, \quad m, n \geq 0$

Answer

(a) The eigenvalues of F_{n+1} are $\{\varphi, \hat{\varphi}\}$, so by Theorem 7.36, the sum $s_n = \sum \binom{n}{j} F_{1+j}$ is a recurrence sequence with the eigenvalues $\{1+\varphi, 1+\hat{\varphi}\} = \{\varphi^2, \hat{\varphi}^2\}$. Therefore, there are constants such that $s_n = c_1 F_{2n} + c_2 F_{2n+1}$. Since $s_0 = 1$and $s_1 = 2$, we see that $c_1 = 0$ and $c_2 = 1$ so $s_n = F_{2n+1}$.

(b) Similarly to part (a), there are constants for which $t_n = \sum \binom{n}{j} F_{2+j} = c_1 F_{2n} + c_2 F_{2n+1}$. Since $t_0 = 1$and $t_1 = 3$, we see that $c_1 = 1$ and $c_2 = 1$ so $t_n = F_{2n} + F_{2n+1} = F_{2n+2}$.

(c) Use induction on m. The result is true for $m = 1$ and $m = 2$ by parts (a) and (b). Now let $m \geq 2$ and assume the result is true for both $m - 1$ and m. Then

$$\sum_j \binom{n}{j} F_{m+1+j} = \sum_j \binom{n}{j} F_{m+j} + \sum_j \binom{n}{j} F_{m-1+j}$$
$$= F_{2n+m} + F_{2n+m-1} = F_{2n+m+1}$$

which shows the result is also true for $m + 1$, completing the induction.

7.4.11. **(a)** Verify that the operator $g(E) = E^2 + E - 1$ annihilates the sequence $(-1)^n F_n$.

(b) Prove that $\sum_{j=0}^{n} (-1)^j F_{j+1} = 1 + (-1)^n F_n$.

Answer

(a) $\left(E^2 + E - 1\right)(-1)^n F_n = (-1)^{n+2} F_{n+2} + (-1)^{n+1} F_{n+1} - (-1)^n F_n$
$= (-1)^n \left(F_{n+2} - F_{n+1} - F_n\right) = 0$

(b) $(E - 1) \sum_{j=0}^{n} (-1)^j F_{j+1} = (-1)^{n+1} F_{n+2}$, so by part (a),

$g(E)(E - 1) \sum_{j=0}^{n} (-1)^j F_{j+1} = g(E)(-1)^{n+1} F_{n+2} = 0$. The spectrum of $g(E)(E - 1)$ is $\{1, -\varphi, -\hat{\varphi}\}$, so there are constants for which

$$\sum_{j=0}^{n} (-1)^j F_{j+1} = c_0 + c_1 (-1)^n F_n + c_2 (-1)^{n+1} F_{n+1}.$$

The first three cases for $n = 0, 1, 2$ give equations for the constants, showing that $c_0 = 1, c_1 = 1, c_2 = 0$. This proves the identity.

7.4.13. Express the sum $t_n = F_0^2 + F_2^2 + F_4^2 + \cdots + F_{2n}^2, n \geq 0$ in closed form.

Answer

The second order operator $\left(E - \varphi^2\right)\left(E - \hat{\varphi}^2\right)$ annihilates the sequence F_{2n}, so the third order operator $f(E) = (E - 1)\left(E - \varphi^4\right)\left(E - \hat{\varphi}^4\right)$ annihilates F_{2n}^2. The operator has the spectrum $\left\{\left(\varphi^2\right)^2, \left(\hat{\varphi}^2\right)^2, \varphi^2\hat{\varphi}^2\right\} = \left\{\varphi^4, \hat{\varphi}^4, 1\right\}$, so F_{2n}^2 is in the vector space spanned by $\left\{1, F_{4n}, F_{4n+1}\right\}$.Since 1 is an eigenvalue of $f(E)$ of multiplicity $r = 1$, Theorem 7.29 tells us that there are constants c_0, c_1, c_2, c_3 for which

$$t_n = F_0^2 + F_2^2 + F_4^2 + \cdots + F_{2n}^2 = c_0 n + c_1 + c_2 F_{4n} + c_3 F_{4n+1}$$

The constants are determined by the values $t_0 = 0, t_1 = 1, t_2 = 10, t_3 = 74$, showing that

$$F_0^2 + F_2^2 + F_4^2 + \cdots + F_{2n}^2 = \frac{1}{5}\left(F_{4n+2} - 2n - 1\right)$$

7.4.15. Express the sum $\sum_{j=0}^{n} j^3$ in closed form.

Answer

The 4th order operator $(E-1)^4$ annihilates the sequence of cubes n^3. Since 1 is an eigenvalue of multiplicity $r = 4$, Theorem 7.29 tells us that there are constants c_0, c_1, c_2, c_3, c_4 for which $\sum_{j=0}^{n} j^3 = c_0 n^4 + c_1 n^3 + c_2 (n+1)^3 + c_3 (n+2)^3 + c_4 (n+3)^3$. A simpler form for this polynomial of degree 4 is $\sum_{j=0}^{n} j^3 = \hat{c}_0 n^4 + \hat{c}_1 n^3 + \hat{c}_2 n^2 + \hat{c}_3 n + \hat{c}_4$. Taking the cases for $n = 0$, 1, 2, 3, 4 determines the coefficients of the polynomial, giving us the sum in a closed form, namely $\sum_{j=0}^{n} j^3 = \frac{1}{4}\left(n^4 + 2n^3 + n^2\right) = \left(\frac{n(n+1)}{2}\right)^2$.

7.4.17. Let s_n be the number of ways that a $1 \times n$ board can be tiled with squares and dominoes, where exactly one red domino is used and any number of dominoes and squares can be used, all of which are white.

(a) Show that $s_n = \sum_{j=1}^{n-1} F_j F_{n-j}$.

(b) Show that $s_n = \frac{1}{5}\left((n-1)F_n + 2nF_{n-1}\right)$ in closed form.

Answer

(a) Suppose that the red domino covers cells j and $j+1$, where $1 \leq j \leq n-1$. There are then $F_j F_{n-j}$ ways to tile the remaining board(s) to the left and/or right of the red domino. Therefore, there are $s_n = \sum_{j=1}^{n-1} F_j F_{n-j}$ ways to tile a board of length $n \geq 2$.

(b) If the Fibonacci operator $f(E) = E^2 - E - 1$ is applied to s_n, we get a sum of terms spanned by the sequences F_n and F_{n+1}. Therefore $f^2(E)s_n = 0$, so s_n is in the vector space spanned by $\{F_n, F_{n+1}, nF_n, nF_{n+1}\}$. That is, s_n has the form $s_n = (c_1 + nc_2)F_{n-1} + (c_3 + nc_4)F_n$. Since $s_2 = 1, s_3 = 2, s_4 = 5, s_5 = 10$, we get the closed expression $s_n = \frac{1}{5}\left((n-1)F_n + 2nF_{n-1}\right)$. The formula also correctly gives $s_1 = 0$.

7.4.19. Let $s_n = \sum \binom{n}{j} F_{2j}$.

(a) Find a recursion relation satisfied by s_n, and use it to compute the first six values of the sequence.

(b) Show that $\sum \binom{n}{j} F_{2j} = \dfrac{\alpha^n - \beta^n}{\sqrt{5}}$, where $\alpha, \beta = \dfrac{5 \pm \sqrt{5}}{2}$.

Answer

(a) Since F_{2n} is annihilated by the polynomial operator $C(E) = (E - \varphi^2)(E - \hat{\varphi}^2) = (E - 1 - \varphi)(E - 1 - \hat{\varphi})$, we know from Theorem 7.36 that s_n is annihilated by

$$\begin{aligned} C(E-1) &= (E - 2 - \varphi)(E - 2 - \hat{\varphi}) = (E - \alpha)(E - \beta) \\ &= E^2 - (\alpha + \beta)E + \alpha\beta = E^2 - 5E + 5 \end{aligned}$$

Therefore, $s_{n+2} = 5s_{n+1} - 5s_n$. Since $s_0 = 0$ and $s_1 = 1$, the sequence begins $0, 1, 5, 20, 75, 275, \ldots$.

(b) There are constants for which $s_n = c_1\alpha^n + c_2\beta^n$, and the initial values show that $s_n = \dfrac{\alpha^n - \beta^n}{\sqrt{5}}$.

7.4.21. It can be proved that $s_n = \sum_j \binom{n}{j-n} P_j$ is a second-order recurrence sequence. Find this recurrence relation using the values $s_0 = 0, s_1 = 3, s_2 = 24, s_3 = 198$.

Answer

The recurrence relation has the form $s_{n+2} = as_{n+1} + bs_n$ so the initial conditions give us the equations $24 = 3a + 0,\ 198 = 24a + 3b$. Solving the linear system shows that $a = 8$ and $b = 2$.

SECTION 7.5

7.5.1. Let $s_n = \sum_j \binom{n}{j-n} h_j$, where h_n is a recurrence sequence of order 2 that satisfies the recurrence relation $h_{n+2} = ah_{n+1} + bh_n$.

(a) Given $f(E) = E^2 - cE - d$, show that

$$f(E)s_n = \sum_j \binom{n}{j-n} \left((a^2 + 2a + b - c + 1)h_{j+2} + (ab + 2b - c)\,h_{j+1} - dh_j\right)$$

(b) Show that $f(E)\,s_n = 0$ if $c = a^2 + a + 2b$ and $d = b(a - b + 1)$.
(c) Prove that $s_{n+2} = \left(a^2 + a + 2b\right) s_{n+1} + \left(b + ab - b^2\right) s_n$.

Answer

(a) First notice that $h_{j+3} = ah_{j+2} + bh_{j+1}$ and $h_{j+4} = ah_{j+3} + bh_{j+2} = a\left(ah_{j+2} + bh_{j+1}\right) + bh_{j+2} = \left(a^2 + b\right) h_{j+2} + abh_{j+1}$. Therefore,

$$
\begin{aligned}
f(E)\,s_n &= \left(E^2 - cE - d\right) s_n \\
&= \sum_j \binom{n+2}{j-n-2} h_j - c\sum_j \binom{n+1}{j-n-1} h_j - d\sum_j \binom{n}{j-n} h_j \\
&= \sum_j \binom{n}{j-n-2} h_j + 2\sum_j \binom{n}{j-n-3} h_j + \sum_j \binom{n}{j-n-4} h_j \\
&\quad - c\sum_j \binom{n}{j-n-1} h_j - c\sum_j \binom{n}{j-n-2} h_j - d\sum_j \binom{n}{j-n} h_j \\
&= \sum_j \binom{n}{j-n} \left(h_{j+4} + 2h_{j+3} + (1-c)h_{j+2} - ch_{j+1} - dh_j\right) \\
&= \sum_j \binom{n}{j-n} \left((a^2 + 2a + b - c + 1)h_{j+2} + (ab + 2b - c)\, h_{j+1} - dh_j\right)
\end{aligned}
$$

(b) For $c = a^2 + a + 2b$ and $d = b + ab - b^2 = b(a - b + 1)$

$$
\begin{aligned}
f(E)\,s_n &= \sum_j \binom{n}{j-n} \left((a^2 + 2a + b - c + 1)h_{j+2} + (ab + 2b - c)\, h_{j+1} - dh_j\right) \\
&= (a - b + 1)\sum_j \binom{n}{j-n} \left(h_{j+2} - ah_{j+1} - bh_j\right) = 0
\end{aligned}
$$

(c) Since $f(E) = E^2 - cE - d = E^2 - \left(a^2 + a + 2b\right) E - \left(b + ab - b^2\right)$ annihilates s_n, the sequence satisfies the recurrence.

7.5.3. Let

$$
s_n = \sum_j \binom{n}{j-n} P_j
$$

where P_n is the Pell sequence. Use the result of Problem 7.5.1 to show that $s_{n+2} - 8s_{n+1} - 2s_n = 0$.

Answer

Since $a = 2$ and $b = 1$, $c = 2^2 + 2 + 2 \cdot 1 = 8$ and $d = 1(2 - 1 + 1) = 2$, s_n is annihilated by the operator $f(E) = E^2 - 8E - 2$. (This verifies Problem 7.4.21.)

7.5.5. Prove that the type of partial fraction decomposition used in the proof of Theorem 7.1 exists for all rational functions for which the degree of the denominator $Q(x)$ exceeds the degree of the numerator $P(x)$.

Answer

Use induction on n, the degree of the denominator. The decomposition exists when $n = 1$, since the rational function is already in the desired form. Now assume the partial fraction decomposition holds for all rational functions whose denominator has degree less than n, where $n \geq 2$. Now consider any rational function whose denominator has degree n and has a root α of multiplicity r. That is, the rational function has the form $\dfrac{P(x)}{(x-\alpha)^r Q(x)}$ where $r \geq 1$, $P(\alpha) \neq 0$, $Q(\alpha) \neq 0$ and degree $P < n$. For any constant c, we have

$$\frac{P(x)}{(x-\alpha)^r Q(x)} = \frac{c}{(x-\alpha)^r} + \frac{P(x) - cQ(x)}{(x-\alpha)^r Q(x)}$$

In particular, for $c = \dfrac{P(\alpha)}{Q(\alpha)}$ we see that $P(\alpha) - cQ(\alpha) = 0$ so $P(x) - cQ(x) = (x-\alpha)\hat{P}(x)$ for some polynomial $\hat{P}(x)$ and $\dfrac{\hat{P}(x)}{(x-\alpha)^{r-1} Q(x)}$ is a rational function with a denominator $\hat{Q}(x) = (x-\alpha)^{r-1} Q(x)$ of degree $n - 1$ and degree $\hat{P} < n - 1$. By the induction hypothesis, $\dfrac{\hat{P}(x)}{\hat{Q}(x)}$ has a partial fraction decomposition of the desired form. When this is added to $\dfrac{P(\alpha)}{Q(\alpha)(x-\alpha)^r}$ the desired partial fraction decomposition is obtained for the given rational function.

8

COUNTING WITH SYMMETRIES

PROBLEM SET 8.2

8.2.1. Let γ and η be the permutations

$$\gamma = \begin{pmatrix} 1 & 2 & 3 & 4 & 5 \\ 4 & 1 & 5 & 2 & 3 \end{pmatrix} \text{ and } \eta = \begin{pmatrix} 1 & 2 & 3 & 4 & 5 \\ 5 & 2 & 4 & 3 & 1 \end{pmatrix}.$$

(a) Write γ and η as products of cycles.

(b) Compute the products $\gamma\eta, \eta\gamma, \gamma^2, \eta^2$, expressing your answers as a product of cycles. [*Note*: $(\gamma\eta)(1) = \gamma(\eta(1)) = \gamma(5) = 3$.]

Answer

(a) $\gamma = (1\ 4\ 2)(3\ 5)$, $\eta = (1\ 5)(2)(3\ 4)$

(b) $\gamma\eta = \begin{pmatrix} 1 & 2 & 3 & 4 & 5 \\ 3 & 1 & 2 & 5 & 4 \end{pmatrix} = (1\ 3\ 2)(4\ 5)$

$$\eta\gamma = \begin{pmatrix} 1 & 2 & 3 & 4 & 5 \\ 3 & 5 & 1 & 2 & 4 \end{pmatrix} = (1\ 3)(2\ 5\ 4)$$

Solutions Manual to Accompany Combinatorial Reasoning: An Introduction to the Art of Counting, First Edition. Duane DeTemple and William Webb.

$$\gamma^2 = \begin{pmatrix} 1 & 2 & 3 & 4 & 5 \\ 2 & 4 & 3 & 1 & 5 \end{pmatrix} = (1\ 2\ 4)\,(3)\,(5)$$

$$\eta^2 = \begin{pmatrix} 1 & 2 & 3 & 4 & 5 \\ 1 & 2 & 3 & 4 & 5 \end{pmatrix} = (1)\,(2)\,(3)\,(4)\,(5)$$

8.2.3. **(a)** What is the order of a k cycle?

(b) What is the order of a product of a 3-cycle and a 4-cycle?

(c) What is the order of a product of a 4-cycle, a 5-cycle, and a 6-cycle?

Answer

(a) k

(b) 12

(c) 60, the least common multiple of 4, 5, and 6

8.2.5. Tabulate all of the products $\gamma\eta$, where $\gamma, \eta \in D_3 = \{e, \rho, \rho^2, \varphi_1, \varphi_2, \varphi_3\}$. For example, since $\rho = (1\ 2\ 3)$ and $\varphi_1 = (1)\,(2\ 3)$ then $\rho\varphi_1 = (1\ 2\ 3)\,(1)\,(2\ 3) = (1\ 2)\,(3) = \varphi_3$.

Answer

	e	ρ	ρ^2	φ_1	φ_2	φ_3
e	e	ρ	ρ^2	φ_1	φ_2	φ_3
ρ	ρ	ρ^2	e	φ_3	φ_1	φ_2
ρ^2	ρ^2	e	ρ	φ_2	φ_3	φ_1
φ_1	φ_1	φ_2	φ_3	e	ρ	ρ^2
φ_2	φ_2	φ_3	φ_1	ρ^2	e	ρ
φ_3	φ_3	φ_1	φ_2	ρ	ρ^2	e

8.2.7. The top of a square table will be covered with nine congruent ceramic tiles, $X = \{x_1, x_2, \ldots, x_9\}$, so its group of symmetries is the cyclic group $G = C_4 = \{e, \rho, \rho^2, \rho^3\}$.

x_4	x_3	x_2
x_5	x_9	x_1
x_6	x_7	x_8

(a) Determine the set of fixed points X_γ for each $\gamma \in G$.

(b) Determine the stabilizer subgroup G_x for each $x \in X$.

(c) Determine the orbits orb(x) for each $x \in X$.

(d) Verify that the length of each orbit is given by $|\text{orb}\,(x)| = \dfrac{|G|}{|G_x|}$.

Answer

(a) $X_e = X, X_\rho = X_{\rho^2} = X_{\rho^3} = \{x_9\}$

(b) $G_{x_i} = \{e\}, i = 1, 2, \ldots, 8$ and $G_{x_9} = C_4$

(c) $\text{orb}\,(x_i) = \{x_1, x_3, x_5, x_7\}, i = 1, 3, 5, 7$
$\text{orb}\,(x_i) = \{x_2, x_4, x_6, x_8\}, i = 2, 4, 6, 8$
$\text{orb}\,(x_9) = \{x_9\}$

(d) $\left|\text{orb}\,(x_i)\right| = \dfrac{|C_4|}{|\{e\}|} = \dfrac{4}{1} = 4, i = 1, 2, \ldots, 8$ and $\left|\text{orb}\,(x_9)\right| = \dfrac{|C_4|}{|C_4|} = \dfrac{4}{4} = 1.$

8.2.9. The triangular table top shown will be tiled with six triangles that are white or blue:

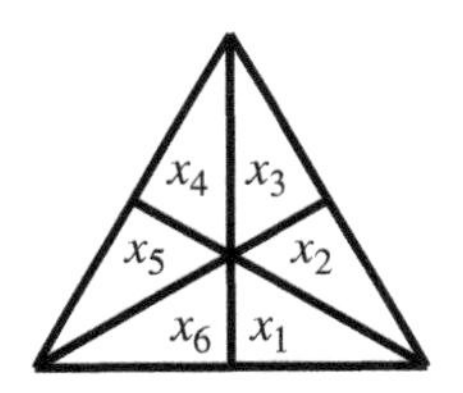

The group of symmetries is $C_3 = \{e, \rho, \rho^2\}$. This induces the permutation group G on [6] for which $e = (1)(2)(3)(4)(5)(6), \rho = (1\ 3\ 5)(2\ 4\ 6)$.

(a) Write ρ^2 as a product of disjoint cycles.

(b) Determine the cycle index of G.

(c) Determine the number of nonequivalent 2-colorings of the triangular table top.

(d) Determine the number of nonequivalent 2-colorings with three tiles of each color.

Answer

(a) $\rho^2 = (1\ 5\ 3)(2\ 6\ 4)$

(b) $Z = \dfrac{1}{3}\left(t_1^6 + 2t_3^2\right)$

(c) $\dfrac{1}{3}\left(2^6 + 2 \cdot 2^2\right) = \dfrac{64 + 8}{3} = 24$

(d) Let $t_1 = w + b$ and $t_3 = w^3 + b^3$ in the cycle index. The coefficient of w^3b^3 is $\dfrac{\binom{6}{3} + 2\,(2)}{3} = \dfrac{24}{3} = 8.$

8.2.11. Write each action $\gamma \in C_4 = \{e, \rho, \rho^2, \rho^3\}$ on the rotationally symmetric arrangement of discs X as a product of disjoint cycles.

Answer

$$e = (1)(2)(3)(4)(5),\ \rho = (1\ 2\ 3\ 4)(5),\ \rho^2 = (13)(24)(5),\ \rho^3 = (1\ 4\ 3\ 2)(5)$$

8.2.13. Compute the stabilizer subgroups G_x of $G = C_4$ for each $x_i \in X$.

Answer

$$G_{x_i} = \{e\} \text{ for } i = 1, 2, 3, 4 \text{ and } G_{x_5} = G = C_4$$

8.2.15. Let $G = D_4 = \{e, \rho, \rho^2, \rho^3, \varphi_1, \varphi_2, \varphi_3, \varphi_4\}$ be the group of actions on the floating pattern of five discs X. Write each action $\gamma \in D_4$ as a product of disjoint cycles. For example, $\varphi_1 = (1)(2\ 4)(3)(5)$.

Answer

$$\begin{aligned} e &= (1)(2)(3)(4)(5),\ \rho = (1\ 2\ 3\ 4)(5),\ \rho^2 = (13)(24)(5),\\ \rho^3 &= (1\ 4\ 3\ 2)(5),\\ \varphi_1 &= (1)(2\ 4)(3)(5),\ \varphi_2 = (1\ 2)(3\ 4)(5),\ \varphi_3 = (1\ 3)(2)(4)(5),\\ \varphi_4 &= (1\ 4)(2\ 3)(5) \end{aligned}$$

8.2.17. Compute the stabilizer subgroups G_x of $G = D_4$ for each $x_i \in X$.

Answer

$$G_{x_1} = G_{x_3} = \{e, \varphi_1\},\ G_{x_2} = G_{x_4} = \{e, \varphi_3\},\ G_{x_5} = D_4$$

8.2.19. **(a)** Compute the cycle index for the cyclic group C_4 acting on X.

(b) Use the cycle index to show there are 12 nonequivalent 2-colorings of the five discs.

(c) Draw a diagram showing the nonequivalent 2-colorings of X.

Answer

(a) $e = (1)(2)(3)(4)(5) \to t_1^5$, $\rho = (1\ 2\ 3\ 4)(5) \to t_1^1t_4^1$, $\rho^2 = (13)(24)(5) \to t_1^1t_2^2$, $\rho^3 = (1\ 4\ 3\ 2)(5) \to t_1^1t_4^1$ and $|C_4| = 4$ so $Z = \frac{1}{4}\left(t_1^5 + 2t_1^1t_4^1 + t_1^1t_2^2\right)$

(b) $\dfrac{2^5 + 2 \cdot 2 \cdot 2 + 2 \cdot 2^2}{4} = \dfrac{32 + 8 + 8}{4} = \dfrac{48}{4} = 12$

(c)

PROBLEM SET 8.3

8.3.1. Let $X = \{x_1, x_2, x_3, x_4, x_5, x_6\}$ be the pattern of six discs shown below, with the symmetries of the dihedral group $D_3 = \{e, \rho, \rho^2, \rho^3, \varphi_1, \varphi_2, \varphi_3\}$, where ρ is a rotation by 120° and the flips φ_i are shown in Example 8.4. The discs have been 2-colored by the function $f = (b, w, b, w, b, w)$.

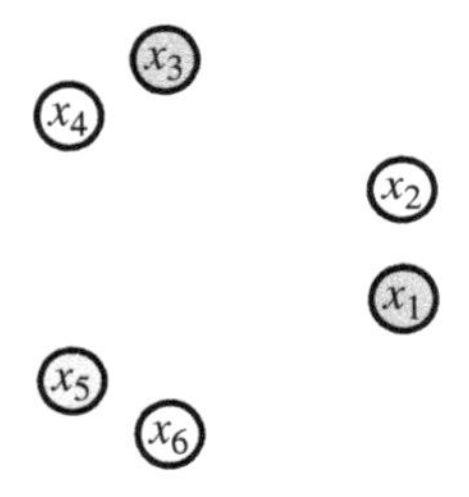

(a) Show that there are three ways to express the orbit $\langle f \rangle$ using different sets of group actions that give the equivalent colorings of f.

(b) Determine the stabilizer subgroup of f.

(c) Use each of the forms of the group actions determining the orbit of f to make a multiplication table with the stabilizer of f.

Answer

(a) $\langle f \rangle = \{ef, \varphi_1 f\} = \{ef, \varphi_2 f\} = \{ef, \varphi_3 f\}$

(b) $G_f = \{e, \rho, \rho^2\}$

(c)

	e	ρ	ρ^2
e	e	ρ	ρ^2
φ_1	φ_1	φ_3	φ_2

	e	ρ	ρ^2
e	e	ρ	ρ^2
φ_2	φ_2	φ_1	φ_3

	e	ρ	ρ^2
e	e	ρ	ρ^2
φ_3	φ_3	φ_2	φ_1

8.3.3. Let $X = \{x_1, x_2, \ldots, x_5\}$ be the vertices of a regular pentagon allowed to float so that its group of actions is induced by the symmetries of the dihedral group D_5. Determine the following sets.

(a) $X_\gamma, \gamma \in D_5$

(b) $G_x, x \in X = \{x_1, x_2, x_3, x_4, x_5\}$

(c) Verify Burnside's Lemma using part (a).

(d) Verify Burnside's Lemma using part (b).

Answer

(a) $X_e = X, X_{\rho^i} = \varnothing, \quad i = 1, 2, 3, 4, \quad X_{\varphi_j} = \{x_j\}, j = 1, 2, 3, 4, 5$

(b) $G_{x_i} = \{e, \varphi_i\}, \quad i = 1, 2, 3, 4, 5$

(c) $\frac{1}{|D_5|} \sum_{\gamma \in D_5} |X_\gamma| = \frac{1}{10}(5 + 0 + 0 + 0 + 0 + 1 + 1 + 1 + 1 + 1) = 1$

(d) $\frac{1}{|D_5|} \sum_{x \in X} |G_x| = \frac{1}{10}(2 + 2 + 2 + 2 + 2) = 1$

8.3.5. Let $X = [24]$ be the set of vertical and horizontal edges that join the vertices of the square grid in the following diagram. Assume that X has the symmetries of the dihedral group $D_4 = \{e, \rho, \rho^2, \rho^3, \rho^4, \varphi_1, \varphi_2, \varphi_3, \varphi_4\}$.

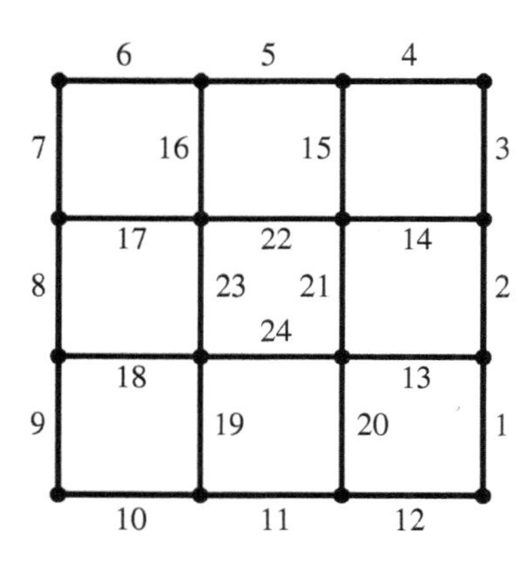

(a) Partition X into equivalence classes.

(b) Verify that the number of equivalence classes obtained in part (a) is given by the formula $k = \frac{1}{|G|} \sum_{\gamma \in G} |X_\gamma|$.

Answer

(a) There are $k = 4$ equivalence classes:
$\{1, 3, 4, 6, 7, 9, 10, 12\}$, $\{2, 5, 8, 11\}$, $\{13, 14, 15, 16, 17, 18, 19, 20\}$, $\{21, 22, 23, 24\}$

(b) The number of edges fixed by each action is given in the list
$|X_e| = |X| = 24, |X_\rho| = |X_{\rho^2}| = |X_{\rho^3}| = 0,$
$|X_{\varphi_1}| = |X_{\varphi_3}| = 4, |X_{\varphi_2}| = |X_{\varphi_4}| = 0$

Therefore $k = \frac{1}{|G|} \sum_{\gamma \in G} |X_\gamma| = \frac{1}{8}(24 + 4 + 4) = \frac{32}{8} = 4.$

8.3.7. A 2×3-*ft* rectangular table is tiled with 1×1*ft* square tiles. Use Burnside's Lemma to determine the number of nonequivalent tables if the tiles are white or blue.

x_3	x_2	x_1
x_4	x_5	x_6

That is, show that the number of nonequivalent 2-colorings is given by each of these formulas.

(a) $k = \frac{1}{|G|} \sum_{\gamma \in G} |T_\gamma|$

(b) $k = \frac{1}{|G|} \sum_{f \in T} |G_f|.$

Answer

The group of actions is induced by the symmetries of the cyclic group $C_2 = \{e, \rho\}$.

(a) $|T_e| = 2^6$ since each of the six tiles can be either of two colors, and $|T_\rho| = 2^3$ since the coloring is determined by the 2^3 ways to color x_1, x_2, x_3. This means there are $k = \frac{1}{2}(2^6 + 2^3) = 36$ nonequivalent two colorings.

(b) There are $2^6 = 64$ colorings of the 6 tiles that are fixed by the identity e. These include $2^3 = 8$ colorings that are also fixed by the rotation (choose a color for each of the pairs (x_1, x_4), (x_2, x_5), and (x_3, x_6)). This means there are $64 - 8 = 56$ colorings fixed only by the identity, for which $|G_f| = |\{e\}| = 1$. For each of the 8 colorings fixed by both the identity and the rotation, we have $|G_f| = |\{e, \rho\}| = 2$. Therefore, there are $\frac{1}{2}(56 \cdot 1 + 8 \cdot 2) = \frac{72}{2} = 36$ nonequivalent colorings.

8.3.9. Complete the following table listing all of the possible monomials for the permutations $\gamma \in S_4$ and the number of permutations with that monomial. Recall that t_j^i corresponds to i cycles of length j.

Monomial	Number of permutations
t_1^4	1
$t_1^2 t_2^1$	6

Answer

Monomial	Number of permutations
t_1^4	1
$t_1^2 t_2^1$	6
$t_1^1 t_3^1$	8
t_4^1	6
t_2^2	3

PROBLEM SET 8.4

8.4.1. Determine the cycle index of the vertices of each of these polygons under rotations.

(a) a regular pentagon

(b) a regular heptagon (7-gon)

(c) a regular hexagon

(d) a regular dodecagon (12-gon)

Answer

(a) $\frac{1}{5}\left(t_1^5 + 4t_5^1\right)$

(b) $\frac{1}{7}\left(t_1^7 + 6t_7^1\right)$

(c) $\frac{1}{6}\left(t_1^6 + t_2^3 + 2t_3^2 + 2t_6^1\right)$

(d) $\frac{1}{12}\left(t_1^{12} + t_2^6 + 2t_3^4 + 2t_4^3 + 2t_6^2 + 4t_{12}^1\right)$

8.4.3. Let $X = \{x_1, x_2, x_3\}$ be the vertices of an equilateral triangle, which is therefore symmetric under the set of actions $\{e, \varphi_1, \varphi_2, \varphi_3\}$ where φ_i is the flip that fixes vertex x_i and exchanges the other two vertices. Is $Z = \frac{1}{4}\left(t_1^3 + 3t_1^1 t_2^1\right)$ the cycle index? If so, the number of 3-colorings is $\frac{1}{4}\left(3^3 + 3 \cdot 3 \cdot 3\right) = \frac{54}{4} = 13\frac{1}{2}$. Explain what is wrong and determine the correct number of 3-colorings of the vertices of the triangle.

Answer

The set $\{e, \varphi_1, \varphi_2, \varphi_3\}$ is not a group since it lacks closure. For example, $\varphi_1\varphi_3 = \rho$, a rotation by 120° is not in the given set. The correct group of the floating triangle is the dihedral group D_3 which has the cycle index $Z = \frac{1}{6}\left(t_1^3 + 2t_3^1 + 3t_1^1t_2^1\right)$ and shows there are $\frac{1}{6}\left(3^3 + 2\cdot 3 + 3\cdot 3\cdot 3\right) = \frac{60}{6} = 10$ nonequivalent 3-colorings.

8.4.5. **(a)** What is the number of nonequivalent 2-colorings of the vertices of a regular hexagon symmetric under rotations?

(b) How many 2-colorings have 3 white and 3 blue vertices?

Answer

(a) The cycle index is $\frac{1}{6}\left(t_1^6 + t_2^3 + 2t_3^2 + 2t_6^1\right)$ so there are $\frac{1}{6}\left(2^6 + 2^3 + 2\cdot 2^2 + 2\cdot 2\right) = 14$ 2-colorings.

(b) The coefficient of w^3b^3 in $\frac{1}{6}\left((w+b)^6 + \left(w^2+b^2\right)^3 + 2\left(w^3+b^3\right)^2 + 2\left(w^6+b^6\right)\right)$ is $\frac{1}{6}\left(\binom{6}{3} + 2\cdot 2\right) = \frac{24}{6} = 4.$

8.4.7. A carbon atom is at the center of a regular tetrahedron, and any combination of hydrogen (H), chlorine (Cl), ethyl (C_2H_5), and methyl (CH_3) can occur at the four vertices $X = \{x_1, x_2, x_3, x_4\}$.

(a) Why is the cycle index for the set of vertices also $Z = \frac{1}{12}\left(t_1^4 + 3t_2^2 + 8t_1^1t_3^1\right)$, the same as for the faces of the tetrahedron as noted in Problem 8.4.6?

(b) What is the number of different nonequivalent molecules?

Answer

(a) The centers of the faces of a regular tetrahedron are the vertices of another regular tetrahedron.

(b) Letting $t_i = 4, i = 1,2,3$ in the cycle index shows that there are $Z(4,4,4) = \frac{1}{12}\left(4^4 + 3\cdot 4^2 + 8\cdot 4\cdot 4\right) = \frac{256+48+128}{12} = 36$ different molecules.

8.4.9. The group of rotational symmetries of the cube was described in Example 8.17, where the cycle index for the set of six faces was derived.

(a) Derive the cycle index for the set of the eight vertices of a cube.

(b) What is the number of nonequivalent 2-colorings of the eight vertices?

(c) How many nonequivalent ways can the eight vertices of a cube be colored with four white and four black vertices?

Answer

(a) The identity has eight 1-cycles, each of the six rotations of type ρ or ρ^3 has two 4-cycles, and each of the three rotations of type ρ^2 has four 2-cycles. Each of the eight rotations of type σ or σ^2 has two 1-cycles and two 3-cycles, and each of the six rotations of type τ has four 2-cycles, so the cycle index is $Z = \frac{1}{24}\left(t_1^8 + 6t_4^2 + 9t_2^4 + 8t_1^2t_3^2\right)$.

(b) $\frac{1}{24}\left(2^8 + 6 \cdot 2^2 + 9 \cdot 2^4 + 8 \cdot 2^2 \cdot 2^2\right) = 23$

(c) The coefficient of w^4b^4 in the pattern inventory is $\frac{1}{24}\left(\binom{8}{4} + 6 \cdot 2 + 9 \cdot \binom{4}{2} + 8 \cdot 4\right) = \frac{168}{24} = 7.$

8.4.11. The following diagram shows two 2-colorings of the vertices of a binary tree on seven vertices.

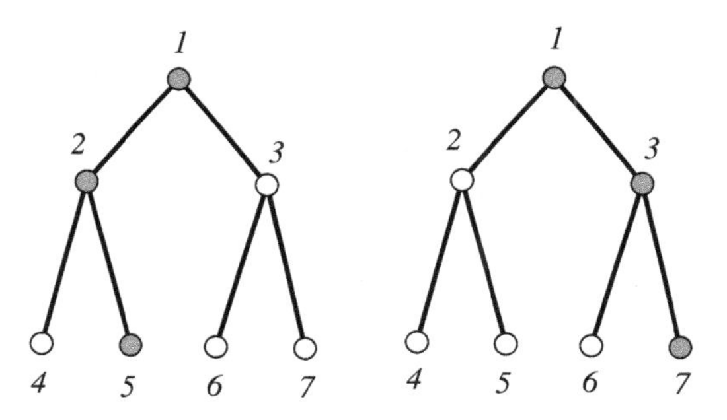

The colorings are equivalent since the direction of branching can be switched to obtain an isomorphic tree. This switching shown in the diagram is given by the action

$$\gamma = \begin{pmatrix} 1 & 2 & 3 & 4 & 5 & 6 & 7 \\ 1 & 3 & 2 & 6 & 7 & 4 & 5 \end{pmatrix} = (1)\,(2\ 3)\,(4\ 6)\,(5\ 7)\,.$$

(a) Derive the group of symmetries of the tree, showing each action as a product of disjoint cycles.

(b) Derive the cycle index for the group.

(c) Derive the 2-color pattern inventory.

(d) In how many nonequivalent ways can the vertices of the tree by colored white or black?

(e) How many 2-colorings of the vertices have four white and three black vertices?

Answer

(a) $G = \{(1)(2)(3)(4)(5)(6)(7), (1)(2\ 3)(4\ 6)(5\ 7),$
$(1)(2\ 3)(4\ 7)(5\ 6), (1)(2)(3)(4\ 5)(6)(7),$
$(1)(2)(3)(4)(5)(6\ 7), (1)(2)(3)(4\ 5)(6\ 7),$
$(1)(2\ 3)(4\ 6\ 5\ 7), (1)(2\ 3)(4\ 7\ 5\ 6)\}$

(b) $Z(G) = \frac{1}{8}\left(t_1^7 + 2t_1^1t_2^3 + 2t_1^5t_2^1 + t_1^3t_2^2 + 2t_1^1t_2^1t_4^1\right)$

(c) $\frac{1}{8}[(w+b)^7 + 2(w+b)\left(w^2+b^2\right)^3 + 2(w+b)^5\left(w^2+b^2\right)$
$+ (w+b)^3\left(w^2+b^2\right)^2 + 2(w+b)\left(w^2+b^2\right)\left(w^4+b^4\right)]$
$= w^7 + 3w^6b + 7w^5b^2 + 10w^4b^3 + 10w^3b^4 + 7w^2b^5 + 3wb^6 + b^7$

(d) $\frac{1}{8}\left(2^7 + 2\cdot 2\cdot 2^3 + 2\cdot 2^5\cdot 2 + 2^3\cdot 2^2 + 2\cdot 2\cdot 2\cdot 2\right) = \frac{336}{8} = 42$

(e) 10

8.4.13. Show that the cycle index of the dihedral group acting on the vertices of a regular m-gon is

$$Z\left(D_m\right) = \begin{cases} \frac{1}{2}Z\left[C_m\right] + \frac{1}{2}t_1^1t_2^{(m-1)/2}, & m \text{ odd} \\ \frac{1}{2}Z\left[C_m\right] + \frac{1}{4}\left(t_2^{m/2} + t_1^2t_2^{(m-2)/2}\right), & m \text{ even} \end{cases}$$

where $Z\left(C_m\right)$ is the cycle index of the cyclic group C_m.

Answer

The sum of monomials due to the rotation actions in D_m are $mZ\left(C_m\right)$. For m odd, each of the m flips has a one 1-cycle and $(m-1)/2$ 2-cycles. This gives the cycle index $Z\left(D_m\right) = \frac{1}{2m}\left(mZ\left(C_m\right) + mt_1^1t_2^{(m-1)/2}\right) = \frac{1}{2}Z\left(C_m\right) + \frac{1}{2}mt_1^1t_2^{(m-1)/2}$, m odd.

For m even, there are $m/2$ flips with $m/2$ 2-cycles and $m/2$ flips with two 1-cycles and $(m-2)/2$ 2-cycles. Therefore the cycle index is $Z\left(D_m\right) = \frac{1}{2m}\left(mZ\left(C_m\right) + \frac{m}{2}t_2^{m/2} + \frac{m}{2}t_1^2t_2^{(m-2)/2}\right) = \frac{1}{2}Z\left(C_m\right) + \frac{1}{4}\left(t_2^{m/2} + t_1^2t_2^{(m-2)/2}\right)$, m even.

PROBLEM SET 8.5

8.5.1. The two permutations of S_2 can be written as a product of transpositions, namely $(1)(2) = (1\ 2)(1\ 2)$ and $(1\ 2)$.

(a) Prove that the six permutations in S_3 can each be written as a product of transpositions. For example, $(1\ 2\ 3) = (1\ 2)(2\ 3)$ and $(1)\ (2\ 3) = (1\ 2)(1\ 2)(2\ 3)$.

(b) Prove that each of the $m!$ permutations of S_m for $m \geq 2$ can be written as a product of transpositions. [*Hint*: Show any cycle can be written as a product of transpositions.]

Answer

(a) The four other permutations can be written $(1)(2)(3) = (1\ 2)(1\ 2)(2\ 3)(2\ 3)$, $(1\ 3\ 2) = (1\ 3)(2\ 3)$, $(2)(1\ 3) = (2\ 3)(2\ 3)(1\ 3)$, and $(1\ 2)(3) = (1\ 2)(2\ 3)(2\ 3)$

(b) Let $\pi = (a\ b\ c\ \ldots\ z)$ be any cycle. Then $\pi = (a\ b)\,(b\ c)\,(c\ d)\cdots(x\ y)\,(y\ z)$.

8.5.3. Verify the following cycle indices.

(a) $Z\left(A_2\right) = t_1^2$

(b) $Z\left(A_3\right) = \frac{1}{3}\left(t_1^3 + 2t_3^1\right)$

(c) $Z\left(A_4\right) = \frac{1}{12}\left(t_1^4 + 8t_1^1 t_3^1 + 3t_2^2\right)$

Answer

(a) $A_2 = \{(1)\,(2)\}$ and $\left|A_2\right| = 1$, so there is one permutation with two 1-cycles.

(b) $A_3 = \{(1)\,(2)\,(3), (1\ 2\ 3), (1\ 3\ 2)\}$ and $\left|A_3\right| = 3$, so there is one permutation with three 1-cycles and two permutations with one 3-cycle.

(c) $A_4 = \left\{\begin{matrix}(1)\,(2)\,(3)\,(4), (1\ 2)\,(3\ 4), (1\ 3)\,(2\ 4), (1\ 4)\,(2\ 3), (1\ 2\ 3)\,(4), (1\ 3\ 2)\,(4), \\ (1\ 3\ 4)\,(2), (1\ 4\ 3)\,(2), (1\ 2\ 4)\,(3), (1\ 4\ 2)\,(3), (1)\,(2\ 3\ 4), (1)\,(2\ 4\ 3)\end{matrix}\right\}$ and $\left|A_4\right| = 4!/2 = 12$ so there is one permutation with four 1-cycles, three with two 2-cycles, and eight with one 1-cycle and one 3-cycle.